PUBLIC CABLE™ 1.0

BUILDING SPECIFICATION SECTION 13410
ELECTRONIC WIRE AND CABLE

A PRO-FORMA BOILERPLATE FOR ENGINEERS,
ARCHITECTS AND SPECIFIERS OF
INTEGRATED SYSTEMS

PART 2 INCLUDES COLLECTED DATA AND BRAND-NEUTRAL
SPECIFICATION DESIGNATION NUMBERS (SDNs) FOR LOW VOLTAGE,
LIMITED ENERGY, POWER LIMITED CABLE TYPES:

COMMUNICATIONS, REMOTE CONTROL, FIRE ALARM, SECURITY,
ACCESS CONTROL, SOUND, AUDIO, VIDEO HD-SDI, SURVEILLANCE CCTV,
BROADBAND CATV, LOCAL AREA NETWORK DATA COMMUNICATIONS, WLAN
WIRELESS 50Ω COAX, BUILDING AUTOMATION AND CONTROL, PUBLIC
ADDRESS, INTERCOM, PAGING, NURSE CALL, MASS NOTIFICATION, SOUND
MASKING, RS-232, RS-422, RS-485 AND CABLES FOR RELATED SYSTEMS

FOR ADOPTION INTO CONSTRUCTION PROJECT SPECIFICATIONS'
APPLICABLE SECTIONS WITHIN

1995 FORMAT	DIVISION 13 AND DIVISION 16
2012 FORMAT	DIVISION 25 AND DIVISION 28

ISICA
INTEGRATED SYSTEMS INFRATRUCTURE CABLING ASSOCIATION

CONSENSUS PUBLISHING
CONGERS, NEW YORK

PUBLIC CABLE™ 1.0

BUILDING SPECIFICATION SECTION 13410
ELECTRONIC WIRE AND CABLE

A PRO-FORMA BOILERPLATE FOR ENGINEERS, ARCHITECTS AND SPECIFIERS OF INTEGRATED SYSTEMS

For an editable document for incorporation into construction project specifications contact the publisher at

Consensus Publishing
75A Lake Road #118
Congers, NY 10920

Edited by Armand Cable for ISICA

ISBN: 978-0-9882364-0-0

LCCN: 2012947842 includes index

Library of Congress subject headings:
Audio-visual equipment - Handbooks, manuals, etc.
TECHNOLOGY & ENGINEERING - Telecommunications.

Buildings -- Specifications.
Specification writing.
Telecommunications wiring.
Buildings - Security measures.
Security systems industry.
Electric wiring.
Systems engineering.
Electronic security systems.
Low voltage systems.
Telecommunication cables.
Fire alarms.
Acoustical engineering.
Sound - Equipment and supplies.
Concurrent Engineering

Buildings - Specifications - United States - Indexes.
Buildings - Specifications - Canada - Indexes.
Telecommunication cables - Design and construction.
Electronic systems - Handbooks, manuals, etc.
Sound - Recording and Reproduction - Handbooks, manuals, etc.
Telecommunication wiring - Handbooks, manuals, etc.
Telecommunication cables - Handbooks, manuals, etc.
Television - Handbooks, manuals, etc.
Digital electronics - Handbooks, manuals, etc.
Television - Equipment and supplies.
Telecommunication - Equipment and supplies.

PUBLIC CABLE™ 1.0

TABLE OF CONTENTS

MASTER TABLE OF PUBLIC CABLE™ 1.0 PRODUCTS

DETAIL TABLE OF PRODUCTS – Stranded Bare Copper, UL minimum Insulation

Solid Bare Copper, UL minimum Insulation

Cabled; Solid Bare Copper; UL Minimum Insulation; **Specification 311**

Multiple Unshielded Pairs; Solid Bare Copper; UL Minimum Insulation; **Specification 312**

Multiple Pairs; EPS; Solid Bare Copper; UL Minimum Insulation; **Specification 314**

Combo Cables; Solid BC; UL Minimum Insul.; Mixed Shielding /AWG Sizes **Specification 315**

Stranded Tinned Copper, Upgrade Insulation

Cabled; Stranded Tinned Copper; Upgrade Insulation; **Specification 321**

Multiple Unshielded Pairs; Stranded Tinned Copper; Upgrade Insulation; **Specification 322**

Multiple Pairs; EPS/CD; Stranded Tinned Copper; Upgrade Insulation; **Specification 323**

Multiple Pairs; EPS; Stranded Tinned Copper; Upgrade Insulation; **Specification 324**

SECTION 13410

ELECTRONIC WIRE AND CABLE

Note to Specification Writers: Some of these products (being Power Limited Cable / Low Voltage cables (rated for up to 150volts or up to 300volts depending) – interconnect systems from Section 16: 16700 Communications and 16810 Sound and Video Circuits. For that reason they are often included in SECTION 16120 CONDUCTORS AND CABLES for Low Voltage Power; less than 600V – which we feel is more expressly for electrical branch circuit wiring. ELECTRONIC WIRE AND CABLE are also frequently specified not in one section but in the several and various sections listed in paragraph 1.2 RELATED SECTIONS.

PART 1 GENERAL

1.1 SECTION INCLUDES

01. Remote Control, Signaling, and Power Limited Circuit Cables, General Purpose
    A. Types CL2 and CL3 commercial cables for Horizontal Use.
    B. Types CL2R and CL3R commercial cables for Riser Use.
    C. Types CL2P and CL3P commercial cables for Plenum Use.

02. Communications Cables, General Purpose
    A. Type CM Communications Cable for Horizontal Use.
    B. Type CMR Communications Cable for Riser Use.
    C. Type CMP Communications Cables for Plenum Use.

03. Cables for Power Limited Fire Alarm Circuits
    A. Type FPL - Power limited fire alarm circuit cable for limited use.
    B. Type FPLR - Power limited fire alarm circuit cable for Riser Use.
    C. Type FPLP - Power limited fire alarm circuit cables for Plenum Use

04. Access Control Cable

05. EIA-232 Cable (RS-232 Cable)

06. EIA-422 Cable (RS-422 Cable)

07. EIA-485 Cable (RS-485 Cable)

08. Category 3 Cable (TIA/EIA-568-C Category 3) voice communication / UTP cable

09. Category 5e Cable (TIA/EIA-568-C Category 5e) copper network cable / UTP cable

10. Category 6 Cable (TIA/EIA-568-C Category 6) copper network cable / UTP cable

11. Surveillance coaxial cable (CCTV coax cable)

12. Broadband coaxial cable (CATV/MATV coax cable)

13. Serial Digital Interface coaxial cable (SDI coax cable)

14. Component RGB and RGBHV high resolution analog and digital video cable (RGB cable)

15. WLAN Wireless transmission coaxial cable (RG-58 coax cable)

1.2 RELATED SECTIONS

01. 13700 Security Access and Surveillance
02 13710 Intrusion Detection
03. 13800 Building Automation and Control
04. 13851 Fire Detection and Alarm
05. 16100 Basic Electrical Materials and Methods
06. 16120 Cables and Conductors (electrical power), Low Voltage, less than 600volts.
07. 16700 Communications
08. 16710 Communications Circuits
09. 16770 Communications Transmission and Reception Equipment
10. 16810 Sound and Video Circuits

1.3 REFERENCES

01. National Fire Protection Association, Inc., NFPA 70: National Electric Code (NEC).
    A. NEC Article 725: Remote Control, Signaling, and Power-Limited Circuits.
    B. NEC Article 760: Fire Protective Signaling Systems.
    C. NEC Article 800: Communications Circuits.
    D. NEC Article 820: Community Antenna Television.
02. National Fire Protection Association, Inc., NFPA 262: Standard Method of Test For Flame Travel and Smoke of Wires and Cables For Use In Air-Handling Spaces. (Formerly UL 910.)
03. CSA C22.1, Canadian Electric Code
04. Underwriter's Laboratory, Inc. (UL):
    A. UL-13. Power Limited Circuit Cables.
    B. UL-444 Communications Cables
    C. UL-1424 Cables for Power-Limited Fire Alarm Circuits
    D. UL-1581 Reference Standard for Electrical Wires, Cables, and Flexible Cords.
    E. UL-1655 Community-Antenna Television Cables
    F. UL-1666 Test for Flame Propagation Height of Electrical and Fiber-Optic Cables Installed Vertically in Shafts.
05. ANSI/TIA/EIA-568-C.0, Generic Telecommunications Cabling Standard, For Customer Premises.
06. ANSI/TIA/EIA-568-C.1, Commercial Building Telecommunications Cabling Standard..
07. ANSI/TIA/EIA-568-C.2, Balanced Twisted Pair Cabling Components.
08. ANSI/TIA/EIA-569, Commercial Building Standard for Telecommunications Pathways and Spaces
09. NECA/BICSI 568, Standard for Installing Commercial Building Telecommunications Systems

1.4 SUBMITTALS

01. Submit under the provisions of Section 01300

A. Before work begins: Manufacturer's Literature or Data Sheet for each Manufacturer and Model Number used in the project. Except: where:
   a. The Manufacturer and Model number appear in Part 2- Products; AND
   b. Manufacturers' Data Sheet unambiguously indicates that Underwriters Laboratory (UL) lists the cable as the Type appropriate for the use (horizontal, riser, or plenum; as applicable.)

then these submittals are waived. If other manufacturers or other model numbers are used; or if Data Sheets indicate a third party other than UL is (or might be) the listing laboratory – submittal of Manufacturers' Data Sheets is required.

B. Before work begins and as technicians are added during work: Installer Certification. Submit for each Installation Technician before he or she participates in any work, a copy of their Certificate to Install Limited Energy Cable or Limited Energy Integrated Systems (or the equivalent) issued by the state authority, if applicable in the state where work is being performed. In addition submit for each Installation Technician a copy of one of the following applicable for the work being performed:
   a. CTS – Certified Technical Specialist (from INFOCOMM)
   b. C-EST – Certified Electronics System Technician (from ESPA - (NSCA / CEA/ CEDIA))
   c. RCDD – Registered Communications Distribution Designer (from BICSI)
   d. NICET or NTC – certification appropriate for the systems being worked on. (National Institute for Certification in Engineering Technologies / National Training Center.)

C. UL Labels from manufacturer's packaging and a list of Model Numbers installed. Two weeks prior to final inspection, submit at least one holographic UL Label pertaining to the flame rating taken from one (at least) package for each Model Number installed. Mark "Label Removed for Submittal" with the date on the package from with label is removed. Submit labels (only one per manufacturer and part number) in association with a list of cables used. Photo images as from as cell phone camera are also acceptable.

D. Certificates: Two weeks prior to final inspection, submit four copies of certification that the material used is in accordance with Part 2 – Products, the drawings and specifications.

E. System labeling schedules, including electronic copy of labeling schedules.

F. Wiring Diagrams. Show typical wiring schematics.

G. Maintenance Data: If applicable

1.5 DEFINITIONS

01. ACR: Attenuation-to-Crosstalk Ratio. ACR requirements are a measurement of signal integrity and quality in certain TIA 568 local area network UTP data transmission cables, and others.

02. AHJ: Authority Having Jurisdiction. Typically the local Building Code or Zoning and Planning Commissioner, Fire Marshall, and any other local authorities having jurisdiction.

03. ANSI: American National Standards Institute.

04. Attenuation: A measure of signal loss over distance. Expressed in decibels (dB) per unit of length. Varies with signal frequency, with higher frequencies attenuating first over distance, and lower frequencies traveling farther (or attenuating less at the same distance.)

05. Audio: Having to do with audible sound communications. Alternatively called Sound. Applicable to Section 16810 Sound and Video circuits.

06. A/V: Audio/Video or Audio/Visual. -- Cables wiring audio distribution devices, audio speakers, and video distribution devices. Interconnecting audio source and audio receiver or video source and video receiver. Having to do with Section 16810 Sound and Video Circuits.

07. AWG: American Wire Gage. The sizing standards for electrically conducting metal wire maintained by various ASTM specifications.

08. AWM: Appliance Wiring Material. A (UL) designation and listing requirement typically for wiring internal to appliances, equipment, machinery, etc. Not typically a requirement for wiring for permanent installation in buildings.

09. Backbone: Cable which is the main -- serving multiple drops in a multi-layer signal distribution system. Alternatively called Trunk Cable.

10. BC: Bare Copper conductor; typically insulated from other cable components. Copper conductor without Tin Coating. Uncoated Copper. Copper manufacture red in accordance with ASTM B3 - 01(2007) Standard Specification for Soft or Annealed Copper Wire (most common).

11. BCB: Bare Copper Braid. A layer of shielding that is a braid of bare copper wire. Typically in CCTV surveillance coaxial cable.

12. BNC: A connector style typically used on Video Coaxial Cable, CCTV Coax, and other types of RF Coaxial Cable. Has a push-twist-pull action for connect / disconnect to / from the mating jack. There are a variety of cable-connection styles. Literally Bayonet Neill-Concelman; named for the inventors.

13. Building Wire: an expression for Electrical wire and cable distributing electricity within buildings including trunk, feeder, and branch circuit wiring. Generally considered to include wire types THHN, XHHW and USE, and cable types SEU, SER, NM-B, and armored cable type AC and metal clad type MC, or their similar Canadian styles where applicable. Generally considered 600V rated but the term is sometimes applied to higher voltages as well. Details within Section 16120.

14. Bunched: A method of cabling where multiple conductors are simply twisted together without a specified conductor sequence ("rotation") or lay-up, and without left-hand lay / right-hand lay / left-hand lay layering. Bunched conductors should rotate counterclockwise (left hand lay.) Bunched conductors should still have a maximum lay length defined as a multiple of the diameter of the insulated wires being bunched (cabled.) More flexible and more easily handled than an assembly of conductors that straight, parallel, and not cabled.

15. Cabled: 1.) Any twisting of multiple conductors including bunched. (Accepted definition for Public Cable.) 2.) ICEA cabling frequently specified for Electric Utility applications; defined as alternating layers of left-hand lay and right-hand lay with the outermost layer being left-hand lay. Also called (erroneously) Planetary Cabling after the cabling equipment required for high conductor count cables. Erroneously (called Planetary Cabling) because not all cable produced on a planetary cabler is necessarily ICEA cabling lay-up compliant. More flexible and more easily handled than an assembly of conductors that straight, parallel, and not cabled.

16. Capacitance: The ability or the tendency of a body to store an electrical charge. An undesirable but unavoidable trait in electronic wire and cable (audio), as it means less crisp signal transmission and causes signal distortion. Capacitance is affected by the insulation material selection and thickness. In audio cable, typically lower capacitance is better. Capacitance is directly and inversely related to impedance. When a system is being designed for a matched end-to-end

characteristic impedance (such as 75 OHM video coax, 50 OHM transmission coax, 100 OHM LAN-data cabling, 100 OHM twisted pair, or 120 OHM EIA 485) – a targeted capacitance can be a desired design feature corresponding to the characteristic impedance, instead of lower being necessarily better. Expressed in pF/ft. (Pico farads per foot.)

17. CAT: Abbreviation of Category.

18. CATV: Community Antenna Television. (1) A means to transmit multiple television channels to consumers through 75-Ohm coaxial cable or optical fiber. (2) The cable over which CATV signals are transmitted. (3) An NRTL listed NEC Article 820 compliant cable having passed the flame test required for horizontal use in premises. Alternatively called Broadband Coaxial Cable or simply Broadband. (4) One of the systems addressed by Section 16710 – Communications Circuits.

19. CATVP: An NRTL listed NEC Article 820 compliant 75-Ohm coaxial cable having passed the flame test required for horizontal use in plenum spaces.

20. CATVR: An NRTL listed NEC Article 820 compliant 75-Ohm coaxial cable having passed the flame test required for horizontal use in risers.

21. CCA: Copper Clad Aluminum. A composite conducting wire type. Properly used in certain 50-Ohm transmission RF coaxial cables as a center conductor. Cable and / or packaging must be clearly marked when used as the braid in CCTV 75-Ohm surveillance coax. Must never to be used as the conductors in TIA-568 local area network cable.

22. CCS: Copper Clad Steel: A composite conducting wire type. Most commonly used as the center conductor in CATV 75-Ohm television signal distribution coaxial cable.

23. CCTV: Closed Circuit Television. (1) One of the systems addressed by Section 13700 Security Access and Surveillance. (2) The 75-Ohm coaxial cable transmitting analog or digital CCTV video. Alternatively called Baseband video.

24. CD: Common Drain (Wire.) In 2-pair EPS cable, when the aluminum foil is facing out on each pair; and a single shared common drain wire drains the noise collected by both shields. (EPS/CD)

25. CEC: Canadian Electrical Code. Published by CSA. Formally CSA C22.1. Also CE code.

26. –CI: Suffix designating listed Circuit Integrity cable type, having passed UL 2196 open cabling testing. e.g. FPLR-CI is a listed Circuit Integrity fire alarm cable.

27. –CIC: Suffix indicating cable that, when installed in a conduit system as approved, completes a (UL) Fire Resistive Directory listed cable-in-conduit (FHIT) system. The cable-in-conduit system represented on the corresponding FHIT has passed UL 2196 circuit integrity in conduit test.

28. CL2: Class 2 signaling. (1) Cable: An NRTL listed NEC Article 725 (CL2) and UL 13 (CL2) compliant cable having passed the flame test required for horizontal use in premises; (UL 1685) - 150V maximum voltage (2) Circuit: Class 2 circuits include wiring for low-energy (100VA or less) and low-voltage (under 30V) loads such as: Security Systems (Section 13700); Limited-Energy Communications Systems including voice (Section 16700 included); Low-Energy Sound Systems (Section 16810 included) such as intercom, paging, nurse-call, sound, and public address systems; also programmable logic controllers (PLC's); thermostats (HVAC) and low-voltage lighting. Additional CL2 use include TIA-568 twisted-pair or coaxial local area networks (LAN) [NEC 2002 725.41(A)(4)]. Since Class 2 circuits limit power to 100VA for circuits operating at 30V or less and 0.5VA for circuits operating 31-150V; protection from electrical fires is thereby accomplished. Risk

of Electric Shock is mitigated by limiting current to 5mA maximum between 30V and 150V [NEC 2002 Chapter 9, Table 11]. Class 2 circuits can be wired with Class 2 (CL2) cable, Class 3 (CL3) cable, Communications Cable (NEC Article 800; CM); substitutes permitted by NEC Articles 725 and 800 [NEC 2002 Table 725.61(A) and others]; depending on the installation (i.e. Riser / Plenum.)

29. CL2P: An NRTL listed NEC Article 725 Class 2 (CL2) and UL 13 (CL2) compliant cable having passed the flame test required for plenum use (NFPA 262 / UL 910) - 150V maximum voltage.

30. CL2R: An NRTL listed NEC Article 725 Class 2 (CL2) and UL 13 (CL2) compliant cable having passed the flame test required for Riser use (UL 1666). 150V maximum voltage.

31. CL3: Class 3 signaling. (1) Cable: An NRTL listed NEC Article 725 (CL3) and UL 13 (CL3) compliant cable having passed the flame test required for horizontal use in premises (UL 1685) - 300V maximum voltage. (2) Circuit: Circuits are Class 3 when the power demand is greater than 0.5VA but is no greater than 100VA for operating voltage above 30V [NEC 2002 Chapter 9, Table 11]. Class 3 signaling circuit applications include for Security Systems (Section 13700); Low-Energy Sound Systems (Section 16810 included) such as intercom, public address,and certain Nurse Call systems; Limited-Energy Communications Systems (Section 16700 included) including voice communication. Permitted levels of voltage and current are higher for Class 3 than for Class 2 circuits. To prevent electric shock hazard, wiring must be rated no less than 300V [NEC 2002 725.71(E) and (F)]. Voltage rating is not printed on the cable however (UL 13.) Cable types and wiring methods that meet Class 3 requirements include NRTL listed PLTC Cable, CL3 Cable, Communications Cable [NEC Article 800; CM); [substitutes listed in NEC 2002 Table 725.61(A) and others], depending on the installation (i.e. Riser / Plenum.)

32. CL3P: An NRTL listed NEC Article 725 Class 3 (CL3) and UL 13 (CL3) compliant cable having passed the flame test required for plenum use (NFPA 262 / UL910). 300V maximum voltage.

33. CL3R: An NRTL listed NEC Article 725 Class 3 (CL3) and UL 13 (CL3) compliant cable having passed the flame test required for Riser use (UL 1666). 300V maximum voltage.

34. Class 2: See CL2

35. Class 3: See CL3

36. CM: Communications Cable. An NRTL listed NEC Article 800 and UL 444 compliant cable having passed the flame test required for horizontal use in premises (UL 1685) - 300V maximum voltage. In Canada: Communications Cable meeting the requirements of CSA C22.2 No. 214/UL 444 and CEC Section 60; and having passed CSA FT1 (UL 1685 (UL 1581, Sec. 1160) Vertical-Tray.)

37. CMG: An NRTL listed CSA C22.2 No. 214/UL 444 and CEC Section 60 compliant Communications Cable having passed CSA FT4 flame test, for permanent installation as permitted

38. CMH: An NRTL listed CSA C22.2 No. 214/UL444 and CEC Section 60 compliant Communications Cable having passed the CSA FT1 flame test. For permanent installation as permitted.

39. CMP: Communications Cable, Plenum use. An NRTL listed NEC Article 800 and UL 444 compliant cable having passed the flame test required for use in plenum spaces (NFPA 262 / UL 910) - 300V maximum voltage. In Canada; An NRTL listed CSA C22.2 No. 214/UL444 and CEC Section 60 compliant cable having passed the CSA FT6 flame test or NFPA 262 (UL 910.)

40. CMR: Communications Cable, Riser use. An NRTL listed NEC Article 800 and UL 444 compliant cable having passed the flame test required for use in risers. (UL 1666) - 300V maximum voltage. . In Canada; An NRTL listed CSA C22.2 No. 214/UL444 and CEC Section 60 compliant cable having passed the CSA FT4 flame test or UL 1666.

41. CMX: Communication Cable meeting flame tested UL 1581, Section 1080 (VW-1) or CSA FT1. Limited to residential use. Cable listed CMX-OUTDOOR is rated for limited outdoor residential use.

42. Coax: Cable constructed of a center conductor, dielectric layer (insulation), and an outer concentric shield (or second conductor sharing the same axis as the center conductor ;) and usually an overall jacket. Sometimes called single conductor cable with a shield. Types include CATV, CCTV, and 50-Ohm RF transmission coax.

43. CSA: Canadian Standards Association. Published CSA C22.1 the Canadian Electrical Code.

44. c(UL): Listed by UL to CSA Safety Standards. i.e. Canadian Standards / UL Listed c(UL)

45. dB: Decibel. A logarithmic measure of signal loss or gain relative to the source level.

46. Dia.: Abbreviation for diameter.

47. Dielectric. An insulating material, usually in reference to coax cable. Alternatively called insulation.

48. Drain Wire: An uninsulated tinned copper wire in continuous contact with an aluminum foil shield to drain away the noise collected.

49. Drop: Cable tapping off a Backbone or Trunk Cable, serving an individual final service point.

50. DW: Drain Wire.

51. EMI: Electro-magnetic interference. (Noise) Particularly lower frequency, higher power interference. (Higher frequency, lower power EMI is commonly called RFI.) Low frequency, high power EMI noise is best addressed with an electrically conductive metallic braid.

52. EPS: Each Pair Shielded. Abbreviation used by several cable manufacturers and here - indicating individually shielded pairs, in paired cable. Same as ISP (Individually Shielded Pairs.) Indicating multiple pairs, each pair being shielded with a foil shield and drain wire.

53. EPS/CD: Each Pair Shielded, Common Drain wire. A two-pair cable made with each pair shielded where the foil is facing out, and a single drain wire drains both foil shields. Also ISP/CD.

54. F: (F Connector. F Type) A connector type typically used on CATV Coaxial Cable with a threaded connect / disconnect action to / from the mating jack. There are a variety of cable-connection styles, compression style being the most preferred.

55. F Type: See F.

56. FAS: CEC and CSA reference; abbreviation for Fire Alarm Signaling (Canadian usage.)

57. FEP: Fluorinated Ethylene-Propylene. An insulating material with high temperature resistance, flame retardancy, and low smoke generation, and excellent resistance to most environmental hazards. Can be foamed. When supplied by DuPont® Corporation – trade name Teflon®.

58. FFEP: Foamed FEP.

59. FHDPE: Foamed High Density Polyethylene.

60. Foam: Material mixed with air or other inert gas. Transmission of signal is improved when insulation is foamed since air is an excellent dielectric and better than the base material. Introducing air into a dielectric improved the properties. Specifically velocity of propagation is increased and capacitance is decreased. Polyethylene (PE) and Fluorinated Ethylene-Propylene (FEP) are some of materials often foamed, creating FPE and FFEP respectively.

61. Foil: A shield layer of aluminum foil reinforced with a laminated layer of polyester or other reinforcement; unless copper foil is described. The presence of a drain wire is necessary and implied; except if a conductive metallic braid shield is present with foil a drain wire is not necessary. Drain wire (or conductive metallic braid) is to be in continuous contact with the foil layer along the entire length of cable.

62. FPE: Foamed PE or Foamed Polyethylene.

63. FPL: Fire Power-Limited Circuit cable. An NRTL listed NEC Article 760 and UL 1424 compliant cable having passed the flame test required for horizontal use in premises (UL 1685) - 300V maximum voltage.

64. FPLP: Fire Power-Limited Circuit cable, Plenum use. An NRTL listed NEC Article 760 and UL 1424 compliant cable having passed the flame test required for use in plenums. (NFPA 262 / UL 910) - 300V maximum voltage.

65. FPLR: Fire Power-Limited Circuit cable, Riser use. An NRTL listed NEC Article 760 and UL 1424 compliant cable having passed the flame test required for use in risers (UL 1666) - 300V maximum voltage.

66. FPLR-CI: Fire Power-Limited Circuit cable, Riser use and Circuit Integrity. An FPLR cable that has additionally passed the 2-Hour circuit integrity test (UL 2196)) – Test for Fire Resistive Cables. FPLR-CI is a (UL) listed cable type.

67. FPLR-CIC: Not a (UL) listed cable type, but reference to cable that is part of a (UL) listed fire resistive Cable-In-Conduit system. Subject cables installed in conduit according to the associated (UL) FHIT conduit system drawing have passed the 2-Hour circuit integrity test (UL 2196) – Test for Fire Resistive Cables; In-Conduit testing.

68. FR: Abbreviation for Flame Retardant (adjective) indicating that a flame retardant has been added to the base material. e.g. FR-PVC or FRPVC is a special formulation of PVC that is more flame retardant than untreated material.

69. Halar®: Solvey Solexis' trademark for ETCFE (a PVDF compound.) A fluoropolymer resisn used as an insulating compound for wire and sometimes as a jacketing material for cable. Like other PVDF compounds, a low-smoke generating material in burn testing used in plenum-rated cables. Cables made with Halar® ETCFE and attributed with 150C temperature rating (UL).

70. Halogen: (Halogens) Five elements, three concerning to wire and cable: Fluorine, Chlorine, and Bromine. Highly corrosive and lethal when released as gas in a fire scenario.

71. HDPE: High Density Polyethylene.

72. Horizontal Use: Cables listed installation in buildings in concealed spaces, but not listed for use in risers and not listed for use in plenum spaces. Cable Types CL2, CL3, CM, CMX (residential use) and FPL (limited use.)

73. IMP: Impedance (abbreviation) Also see OHMs.

74. Impedance: (1) Total opposition that a circuit offers to the flow of a varying current (signal.) (2) Characteristic or design impedance of a cable or connector expressed in Ohms.

75. INS: Insulation (abbreviation.)

76. ISP: Individually Shielded Pairs. Same as EPS. Indicating multiple pairs, each pair being shielded with a foil shield and drain wire.

77. ISP/CD: Individually Shielded Pairs, Common Drain wire. A two-pair cable made with each pair shielded; where the foils are facing out. A single drain wire drains both foil shields. Also EPS/CD.

78. JKT: Jacket (abbreviation.) Outermost layer of a cable containing multiple components.

79. LAN: Local Area Network.

80. Lay: In twisted or cabled conductors, the length of one complete twist. Expressed in inches or twists per foot. (3in lay = 4 twists per foot.) Good cabling practice prescribes a maximum lay as a multiple of the diameter of the insulated wires being twisted or cabled. Direction of the outermost layer is left-hand lay (counter clockwise) for good cabling practice, including where alternating layers of left-hand lay and right hand lay is being observed.

81. LDPE: Low Density Polyethylene.

82. LHL: Left-Hand-Lay. Counterclockwise rotation in cabled conductors.

83. Limited Use: For use as open wiring in residential use only. In commercial construction, may be installed in conduit and up to10 feet may be exposed.

84. LS-PVC: Low Smoke PVC. A formulation of PVC compound that smokes less and with optical clarity such that it is useful in manufacturing cables that pass NFPA 262 for use in plenum spaces. Compounds for insulating material and compounds for jacketing material deemed Low Smoke.

85. LSZH: Low Smoke Zero Halogen. Various standards and industry conventions are cited for materials and for completed cables for: a) Low Smoke properties and b) halogen content.

86. MATV: Master Antenna Television (see also CATV.)

87. Mft: Abbreviation for 1000 feet.

88. Mylar®: DuPont trademark for polyester. A material used to make tapes: used as binder tape over cable cores, to maintain cable integrity after cabling, before jacketing; or as a separator to facilitate jacket removal for conductor termination and final installation; or as a reinforcing layer in laminated shielding tape such as Aluminum-Polyester (Aluminum-Mylar) shielding tape.

89. NEC: National Electrical Code (USA.) Published by the NFPA. Formally NFPA-70.

90. NFPA: National Fire Protection Association. Publisher of NFPA-70 (The National Electrical Code (NEC) and other publications with the shared mission of fire safety and public safety.

91. NPLF: Non-Power Limited Fire Alarm cable. For fire alarm circuits not protected by a classified power limiter. Manufactured and tested in accordance with ANSI/UL 1425, and listed by UL. NPLF (UL Listed) may be used in buildings in horizontal runs. NPLFR (UL Listed) may be used in riser spaces. NPLFP (UL Listed) may be used in plenum spaces.

92. NRTL: Nationally Recognized Test Laboratory per OSHA.

93. N/S: Non-Shielded. Same as NS; Unshielded; Unsh. ; Unshld. Cable is assumed to be unshielded when shielding is not specified and not addressed in a product description.

94. OAS: Overall Shield. Same as OS or simple "Shielded" in a description of multiconductor cable. In a multi-pair cable; OAS means pairs are not shielded individually (unless stated), but shielded overall.

95. OFC: Oxygen Free Copper. Copper manufactured in accordance with the requirements of ASTM B170 - 99(2010)e1: Standard Specification for Oxygen-Free Electrolytic Copper—Refinery Shapes. OFC exhibits slightly superior performance in heat-stress scenarios and slightly better work-hardening and embrittlement characteristics than other copper types. Although OFC has no quantifiable or proven electrical or electronic conductivity advantages, some audio enthusiasts specify it - particularly for speaker cable.

96. OHM: (Also "Ω" – (omega)) 1.) Units of electrical resistance of an electrical conductor, expressed as DC resistance (in OHMs) per 1000ft at a temperature. 2.) Units of characteristic Impedance of a cable, connector, or circuit. Common cable impedances are 75Ω Coaxial Cable (Video, CATV, CCTV, Broadband); 50Ω Coaxial Cable (Wireless transmission, radio and over-the-air television transmission between antennas and transmit / receive equipment); 100Ω twisted pair (Ethernet, LAN data cable, and other computer interconnect protocols); 120Ω twisted pair (EIA-485, DMX-512, and other computer interconnect protocols.)

97. OSP: Outside Plant. Referring to cables for installation outside of the climate controlled building envelope, usually telecommunications cables.

98. PE: Polyethylene. A thermoplastic material with excellent resistance to moisture and sunlight and many other environmental hazards, used as cable insulation and also as a cable jacket material. Polyethylene insulation has lower capacitance than PVC insulation and thus provides improved signal transmission. Polyethylene can be colored and is relatively economical but is not inherently flame retardant.

99. pF/ft: (Pico farads per foot, "puffs" per foot, "puffs") Units of cable capacitance.

100. Plenum, Riser, or Tray Use (installations.) Raceways and spaces in which cables may be routed without conduit. Cables are required to have the appropriate NRTL listing for these uses.

101. Plenum Space: Concealed spaces through which HVAC system draws return air, particularly in the absence of Return Air Duct work. Commonly the space channeling HVAC return air above a drop ceiling and below a raised floor when no return air ductwork is present.

102. Plenum Use: Cable installation without conduit or other raceway in plenum spaces, also called air handling spaces. Cables are required to have a listing from UL of CL2P, CL3P, CMP, or FPLP. Cables with these ratings are also suitable for Riser and Horizontal Uses but are not suitable for use in cable tray unless further listed as such.

103. PP: Polypropylene. A thermoplastic material used as insulation. Polypropylene insulation has lower capacitance than PVC insulation and thus provides improved signal transmission. Polypropylene can be colored although the resulting color is not as vivid and bright as with PVC.

104. PR: Pair. Two twisted insulated wires. Pairing (twisting) provides a measure of common mode noise rejection since current induced by the outside environmental noise ultimately cancels out in the two wires, due to their twisting through the offending noise field. Also lower case "pr."

105. Public Cable: A term for electronic wire and cable distributing low voltage and low energy signals within and between buildings. Generally considered to include cable types CL2, CL3, CM, FPL, and CATV and their more flame retardant variations for use in risers or plenum spaces. So named because the services distributed serve the public safeguarding them, instructing them, directing them, or informing, entertaining, and educating them. Services distributed include but are not limited to security including access control, closed circuit surveillance video, community antenna television, high definition video, component video, audio, life safety circuits including fire alarm circuits, sound systems including mass notification, public address, nurse-call, intercom, paging, and other speaker cable services, remote control and signaling circuits, communications and Ethernet LAN data circuits, sonic sound suppression, etc. Detailed collectively in section 13410 or separately in the various sections assigned to specific services.

106. PVC: Polyvinyl chloride. A versatile thermoplastic insulation and jacket material, available in a wide variety of grades with a wide variety of features and benefits.

107. PVDF: Polyvinylidene fluoride, or polyvinylidene difluoride A fluoropolymer resisn used as an insulating compound for wire and sometimes as a jacketing material for cable. PVDF compounds are flame retardant and low-smoke generating in burn testing and so are useful for plenum-rated cables. Cables made with PVDF jackets and insulation are typically attributed with 150C temperature rating. Also see Halar® and Solef®.

108. RF: Radio Frequency. A general term for signals transmitted over coax; typically 50-Ohm for wireless and 2-way radio; or 75-Ohm for video and other communications.

109. RF Transmission Coax: 50-Ohm coax for transmitting signals between antennas and the transmitting/receiving device, such as radios, (land mobile, citizen band, police / fire / ambulance wireless bands, etc.) Wi-Fi Access Points, and Cellular Telephone Service equipment, wireless microphone, etc.

110. RFI: Radio Frequency Interference. High frequency electromagnetic interference (noise.) Best addressed with 100% coverage foil shielding and drain wire(s). Sources can be transmitting radio antennas, natural background noise such as the sun and natural or unpowered man-made objects.

111. RGB: 1.) A type of component video transmission in which the RED, GREEN, and BLUE color components of the video signal are transmitted separately between the video source and receiver. 2.) A cable comprised of multiple coaxial cables, typically miniature in size, used for the transmission of RGB signals. Also RGBS (4 components) or RGBHV (5 components) for use depending on how the horizontal hold and vertical synchronization, and other embedded signals are processed by the source and receiving equipment.

112. Riser Space: A vertical shaft through one or more floors without fire stopping.

113. Riser Use: Cable installation without conduit or other raceway in riser spaces. Cables that have a UL listing of CL2R, CL3R, CMR, or FPLR. Cables with these ratings are also suitable for Horizontal Uses, but are not suitable for use in Plenum Spaces or for use in cable tray unless further listed as such.

114. SBC: Solid Bare Copper

115. ScTP: A European reference meaning shielded category cable. Screened Twisted Pair. Usually a LAN / Data Cat 5e or Cat 6 cable, unshielded twisted pairs (UTP) with an overall shield (OAS.) "Screened" is a European reference meaning shielded. Same as Shielded Category Cable.

116. Solef®: A Solvey trademark for one of their formulations of PVDF. A fluoropolymer resisn used as an insulating compound for wire and sometimes as a jacketing material for cable. Like other PVDF compounds, Solef is a flame retardant and low-smoke generating material useful in plenum-rated cables. Cables made with Solef insulation and jacket can be attributed with 150C temperature rating (UL).

117. Shield: Conductive layer positioned around center conductor (in coax); twisted conductors (pairs) and/or around overall cable core and under the cable jacket. The shield isolates the cable from environmental electromagnetic interference (EMI) and/or radio frequency interference (RFI) and vise verse. Shielding guards both noise ingress and egress. Common shielding materials are aluminum foil (laminated tape with a layer of reinforcing polyester) with a tinned copper drain wire; a braid of aluminum wire, bare copper wire, or tinned copper wire, or "Foil Plus Braid": aluminum foil (laminated tape with a layer of reinforcing polyester) with a braid tinned copper (paired data grade cables) or with a braid of aluminum wires (CATV broadband coax.) Foil layers and braids collect noise from around the entire circumference. Braids and drain wires provide low electrical resistance and provide longitudinal path to ground for the collected noise.

118. SMPTE: Society of Motion Picture Television Engineers.

119. SOL: Solid conductor.

120. STP: Shielded Twisted Pair or Pairs. Abbreviation used by more than one cable manufacturer indicating shielding; usually individually shielded pairs (ISP) and being the same as each pair shielded (EPS.) Indicates multiple pairs, each pair being shielded with a foil shield and drain wire. Can be misunderstood for twisted pairs with an overall shield.

121. STR: Stranded conductor, more flexible and with greater flex life than solid conductor of the same AWG size.

122. TC 1.) Tray Cable – 600V Cable intended for use in cable tray installations. UL 1277 and / or various ICEA specifications are pertinent and applicable guidelines. 2.) Tinned Copper: manufactured in compliance with ASTM B33 – 10: Standard Specification for Tin-Coated Soft or Annealed Copper Wire for Electrical Purposes. Tinned Copper: see also'TD'

123. TCB: Tinned Copper Braid. – a shield layer, usually in conjunction with a reinforced aluminum foil tape layer.

124. TD: Tinned Copper: manufactured in compliance with ASTM B33 – 10: Standard Specification for Tin-Coated Soft or Annealed Copper Wire for Electrical Purposes.

125. Teflon®: DuPont trademark for FEP.

126. Tray Use: Cable listed for use in cable trays. Cables are required to have a listing from UL of Type PLTC or Type TC. Also called Cable Tray Use.

127. Trunk Cable. See Backbone.

128. TSP: Twisted Shielded Pair. Seldom used abbreviation. Same as EPS, STP, and ISP.

129. (UL): Underwriters Laboratories' registered trademark.

130. UTP: Unshielded Twisted Pairs. Abbreviation usually implying LAN Data TIA-568 Category 5e or Category 6 cable.

131. Video: Having to do with transmission of picture content and usually motion picture media. Most protocol specifications being SMPTE standards.

132. VOP: Velocity of Propagation. Velocity of signal transmission, expressed as a percentage. This is a function of the insulation (dielectric) material selected, and if air is blended into the dielectric (foamed.) 100% would be the speed of light in a vacuum. VOP typically ranges from approximately 55% for PVC to 84% for highly foamed polyethylene or FEP dielectrics. Higher velocity of propagation results in lower loss, and therefore longer distance capabilities, before signal attenuates to unrecognizable levels.

133. WAN: Wide Area Network

134. WBT: Waterblocking Tape. A tape layer that swells when wetted, in turn blocking the ingress of water into the cable. Cables are more likely to then dry out after a storm scenario and remain useful, rather than necessarily and automatically being replaced.

1.6 QUALITY ASSURANCE

01. Manufacturer Qualifications: Minimum of 10 years experience manufacturing products specified in this Section. Manufacturer will have been listed by the NRTL for at least 10 uninterrupted years, including being subjected to the routine on-site follow-up services required to maintain such listing. The frequency of routine follow-up on-site listing maintenance visits shall be production volume based, not calendar based. If a wire type is new (less than 10 years since definition) or new to an otherwise established manufacturer, the 10 years requirement may be waived.

02. Installer Qualifications: Minimum of 2 years experience installing products specified in this Section. Installers will be certified in accordance with Paragraph 1.4 A 2 above, and certification submitted as required there.

03. Product Requirements:

A. Product manufacturer shall be of high reputation, well established in the items particular to the project, distributed nationally, and available from a local distributor known to project management.
B. Product manufacturer and model number will be selected from Paragraph 2.14.
C. Product shall be free from defects in material or workmanship.
D. All product permanently installed will be new, and in its original factory packaging. Any product re-packaged or cut-to-order-length before arrival on the jobsite will be work performed by a UL certified distributor with 5 years minimum experience in such re-packaging and cutting.

E. All product permanently installed will be new, and in its original factory packaging. Any product re-packaged or cut-to-order-length before arrival on the jobsite will be work performed by a UL certified distributor with 5 years minimum experience in such re-packaging and cutting.

F. Product shall be lot-traceable by date code and factory manufacturing order or other unique lot control number.

04. NRTL Requirements:

A. Wire Type, Flame Test Listing NRTL: Only a listing for the NEC wire-type is required. This listing will by definition incorporate the life-safety flame-spread / flame test listing. The NRTL listing all wire-types of all manufacturers of any permanently installed Electronic Wire used in the project shall:

1. Have been listing cables of the type for at least ten years, or since the inception of the cable type if less than 10 years.
2. Conduct regular follow-up service with manufacturers listed at the manufacturer's factory. The frequency of follow-up inspection visits will be based on footage produced and/or labels sold, not on calendar days.
3. Write and maintain industry recognized standard specifications concerning more than one the subject wire types. Examples applicable to Section 13410 are: UL 13 (Power Limited Circuit Cables), UL 444 (Communication Cables), UL 1424 (Cables for Power-Limited Fire-Alarm Circuits), UL 1425, (Cables for Non-Power Limited Fire-Alarm Circuits).
4. Write and maintain industry recognized standard specifications for the flame spread and other burn characteristics; including test methods and pass/fail criteria; for cable types permitted for Horizontal Use, for Riser Use, for Use in Plenum Spaces (or be a historical contributor), and for Fire-Resistive Cables. Examples applicable to Section 13410 are: UL 1581 (Reference Standard for Electrical Wires, Cables, and Portable Cords), UL 1685 (Vertical-Tray Fire-Propagation and Smoke-Release Test for Electrical and Fiber-Optic Cables), UL 1666 (Test for Flame Propagation Height of Electrical and Optical-Fiber Cables Installed Vertically in Shafts), NFPA 262 also referred to as UL 910 and Steiner Tunnel Test (Standard Method of Test for Flame Travel and Smoke of Wires and Cables for Use In Air-Handling Spaces), UL 2196 (Tests for Fire Resistive Cables.)
5. Be common to three or more major manufacturers and/or distributors of the cable type.
6. Be acceptable to New York City as eligible to list cables compliant with Local Law 5; fire alarm circuit cables for installation in the jurisdiction.
7. Maintain laboratory facilities for testing flame and smoke generation of cable under fire conditions for all the tests mentioned in 4 above, including UL 2196 testing.
8. Wire type and flame test compliance listing will be maintained by (UL).
9. Other NRTL's for wire type and flame test compliance will not be permitted.

B. Optional NRTL for electronic features and characterization: 3rd party testing and certification of electronics features and characterization are optional, and different NRTL's are acceptable. Manufacturer self-certification is also acceptable as long as:

1. The manufacturer, make and model (or similar model) have been successfully deployed in similar systems previously.
2. The manufacturer is of good reputation for cables with these electronic features. The manufacturer is handled by a local distributor supplying several cable types from this manufacturer, to the project in question.
3. The manufacturer claims are accepted by project management and the supplying distributor.

Such electronic features and characteristics being (but not limited to) manufacturer claims of compliance with the electrical and electronic features defined by and for :

a) TIA-568 requirements for Category 5e, and Category 6 LAN data communications cables.
b) EIA-232, EIA-422, or EIA-485 compliance
c) SCTE requirements for CATV 75 Ohm coaxial cable.
d) SMPTE requirements for SDI digital video coaxial cable and RGB component video cable.
e) Any other claims of specific impedance, capacitance, crosstalk of various measure between various cables or components of cables, velocity of propagation, attenuation at various frequencies, VSWR (voltage standing wave ratio,) RL (return loss,) etc. etc. etc. – that manufacturers may claim or that may be recommended by stake holders of the integrated systems being installed. Such values may be expressed as minimum, maximum, or nominal, or as may be customary for the attribute; and may be empirical or measured.

1.7 DELIVERY, STORAGE, AND HANDLING

01. Inspect product upon receipt at Jobsite:
    A. Ensure packaging is factory original; or as expected if pre-arranged otherwise. Cartons should be undamaged and unopened. Reels should be full lengths as packaged at the factory with reel-wrap that has not been disturbed.
    B. Collect one holographic UL label from each cable type for submittal per paragraph 1.4 01. C. A photo as from a cell phone camera is also acceptable for submittal purposes.

02. Store indoors, in a dry location maintained above 32 degree F: If cable is stored where the temperature falls below 32 degrees F move the cables to a dry location where the temperature is above 32 degrees F and leave them there for 24 hours to condition prior to opening, handling, or installing. If installation will be conducted where the temperature is below 32 degrees F, ensure they have been permitted to condition for 24 hours in a dry location maintained at 32F or higher.

03. Never apply paint over wire and cable: Field application of paint on wires and cables shall void the manufacturer's warranty and invalidate the flame retardancy and any smoke generation listings

1.8 PROJECT CONDITIONS:

01. Maintain temperature above 32F, and normal humidity. If project temperature falls below 32F, condition cable for 24 hours at 32F or higher. Do not install products at temperatures above or below the temperature limits customary for cables of the type, or outside the limits given by the cable manufacturer.

1.9 WARRANTY

01. Product is warranted to be free of defects in material or workmanship.

02. Product is warranted to meet manufacturer's published specification and perform within the design limits.

03. Field application of paint on wires and cables shall void the manufacturer's warranty and will invalidate the flame retardancy and any smoke generation listings.

PART 2 PRODUCTS

2.1 MANUFACTURERS:

01. The manufacturers listed are acceptable when:
    A. Product data sheets indicate the product is absolutely UL listed for UL cable types or c(UL) listed for CSA cable types regarding the applicable flame test and wire type, and
    B. Product received at the jobsite has the required holographic UL / c(UL) labels.

02. Acceptable Manufacturers:

    A. Tappan Wire & Cable, which is located at: 100 Bradley Parkway; Blauvelt, NY 10913; Toll Free Tel. 1-800-937-3411; Telephone. 1-845-353-9000; Fax 1-845-353-9315; email: salesdept@tappanwire.com; Web: www.tappanwire.com

    B. General Cable, which is located at: 4 Tessenneer Drive, Highland Height, KY 41076; Tel: 859-572-8000; Fax: 859-572-8458; Email: info@generalcable.com Web: www.generalcable.com

    C. West Penn Wire, which is located at: 2833 West Chestnut Street, Washington, Pennsylvania, 15301; Toll Free Tel. 1-800-245-4964; Telephone. 1-724-222-7060; Fax 1-724-222-6420; email: info@westpenn-wpw.com, Web: www.westpenn-wpw.com

    D. Belden, which is located at: 401 Pennsylvania Parkway #200; Indianapolis, IN 46290; Toll Free Tel. 1-800-235-3361; Telephone. 1-317-818-6300; Fax 1-317-818-6382; email: info@belden.com; Web: www.belden.com

2.2 LOW VOLTAGE SECURITY, AUDIO, AND CONTROL CABLING

01. Cable for these sections and systems:

13700	Security portion
13710	Intrusion Detection
13800	Building Automation and Control
16710	Communications Circuits – multi-conductor and paired cable types
16740	Communications and Data Processing Devices
16810	Sound and Video Circuits
16820	Sound Reinforcement
16821	Public Address and Music Equipment
16822	Sound Masking Systems
16830	Broadcast Studio Audio Equipment
16840	Broadcast Studio Video Equipment
16850	Television Equipment
16880	Multimedia Equipment

Intercom, Paging, Nurse Call, Mass Notification and other systems of this nature.

Cables of gage size and construction details called for on project drawings or project specifications shall be selected from the cable specifications and tables under 2.14.

Note to Specification Writers:
LOW VOLTAGE SECUIRTY, AUDIO, AND CONTROL CABLING are typically selected from tables:

- 31UR – Stranded Bare Copper, unshielded multi-conductor, riser rated.
- 31UP – Stranded Bare Copper, unshielded multi-conductor, plenum rated.
- 31SR – Stranded Bare Copper, shielded multi-conductor, riser rated.
- 31SP – Stranded Bare Copper, shielded multi-conductor, plenum rated.
- 32UR – Stranded Bare Copper, unshielded multi-pair, riser rated.
- 32UP – Stranded Bare Copper, unshielded multi-pair, plenum rated.
- 32SR – Stranded Bare Copper, overall shielded multi-pair, riser rated.
- 32SP – Stranded Bare Copper, overall shielded multi-pair, plenum rated
- 33UR – Stranded Bare Copper, each pair shielded multi-pair, common drain wire, riser rated
- 33UP – Stranded Bare Copper, each pair shielded multi-pair, common drain wire, plenum rated
- 34UR – Stranded Bare Copper, each pair shielded multi-pair, riser rated
- 34UP – Stranded Bare Copper, each pair shielded multi-pair, plenum rated
- 35MR – Stranded Bare Copper, mixed shielding and/or gage sizes; riser rated
- 35MP – Stranded Bare Copper, mixed shielding and/or gage sizes, plenum rated
- 36MR – Access Control Composite Cable, Card Reader only shielded; riser rated
- 36MP – Access Control Composite Cable, Card Reader only shielded; plenum rated
- 36SR – Access Control Composite Cable, all functions shielded; riser rated
- 36SP – Access Control Composite Cable, all functions shielded; plenum rated
- 27UH – Hi-Strand Audio & Speaker Cable, In-Wall rated, horizontal rated
- 27UP – Stranded Bare Copper Audio & Speaker Cable, plenum rated
- 311UR – Solid Bare Copper, unshielded multi-conductor, riser rated.
- 311UP – Solid Bare Copper, unshielded multi-conductor, plenum rated.
- 311SR – Solid Bare Copper, shielded multi-conductor, riser rated.
- 311SP – Solid Bare Copper, shielded multi-conductor, plenum rated.
- 312UR – Solid Bare Copper, unshielded multi-pair, riser rated.
- 312UP – Solid Bare Copper, unshielded multi-pair, plenum rated.
- 312SR – Solid Bare Copper, overall shielded multi-pair, riser rated.
- 312SP – Solid Bare Copper, overall shielded multi-pair, plenum rated
- 314UR – Solid Bare Copper, each pair shielded multi-pair, riser rated
- 314UP – Solid Bare Copper, each pair shielded multi-pair, plenum rated
- 315MR – Solid Bare Copper, mixed shielding and/or gage sizes; riser rated
- 315MP – Solid Bare Copper, mixed shielding and/or gage sizes, plenum rated
- 321UH – Stranded Tinned Copper, Upgrade Insulation; unshielded multi-conductor, horizontal rated.
- 321UP – Stranded Tinned Copper, Upgrade Insulation; unshielded multi-conductor, plenum rated.
- 321SH – Stranded Tinned Copper, Upgrade Insulation; shielded multi-conductor, horizontal rated.
- 321SP – Stranded Tinned Copper, Upgrade Insulation; shielded multi-conductor, plenum rated.
- 322UH – Stranded Tinned Copper, Upgrade Insulation; unshielded multi-pair, horizontal rated.
- 322UP – Stranded Tinned Copper, Upgrade Insulation; unshielded multi-pair, plenum rated.
- 322SH – Stranded Tinned Copper, Upgrade Insulation; overall shielded multi-pair, horizontal rated.
- 322SP – Stranded Tinned Copper, Upgrade Insulation; overall shielded multi-pair, plenum rated
- 323UH – Stranded TC, Upgrade Insulation; each pair shielded; multi-pair, common drain, horizontal rated
- 323UP – Stranded TC, Upgrade Insulation; each pair shld; multi-pair; common drain wire, plenum rated
- 324UH – Stranded Tinned Copper, Upgrade Insulation; each pair shielded multi-pair, horizontal rated
- 324UP – Stranded Tinned Copper, Upgrade Insulation; each pair shielded multi-pair, plenum rated

2.3 ACCESS CONTROL CABLING

01. Cable for these sections and systems:

13700 ACCESS CONTROL portion;

Cables of gage sizes and construction details called for on project drawings or project specifications shall be selected from the cable specifications and tables under 2.14.

> Note to Specification Writers:
> ACCESS CONTROL CABLING are typically selected from tables identified above in 2.2 01. for LOW VOLTAGE SECUIRTY, AUDIO, AND CONTROL CABLING but especially from these in particular:
> 35MR – Access Control Combination Cable Bundle, mixed AWG sizes, card reader shielded, riser rated
> 36MP – Access Control Combination Cable Bundle, mixed AWG sizes, card reader shielded, plenum rated
> 36SR – Access Control Combination Cable Bundle, mixed AWG sizes, all functions shielded, riser rated
> 36SP – Access Control Combination Cable Bundle, mixed AWG sizes, all functions shielded, plenum rated

2.4 FIRE DETECTION AND ALARM CABLING

01. Cable for these sections and systems:
13851 Fire Detection and Alarm

Cables of gage size and construction details called for on project drawings or project specifications shall be selected from the cable specifications and tables under 2.14.

> Note to Specification Writers:
> FIRE DETECTION AND ALARM CABLING are typically selected from tables:
> 341UH – Solid Bare Copper, unshielded multi-conductor, horizontal or riser rated FPL and FPLR.
> 341UP – Solid Bare Copper, unshielded multi-conductor, plenum rated FPLP.
> 341SH – Solid Bare Copper, shielded multi-conductor, horizontal or riser rated FPL and FPLR.
> 341SP – Solid Bare Copper, shielded multi-conductor, plenum rated FPLP.
> 342UR – Stranded Bare Copper, unshielded multi-conductor, riser rated FPLR.
> 342UP – Stranded Bare Copper, unshielded multi-conductor, plenum rated FPLP.
> 342SR – Stranded Bare Copper, overall shielded multi-conductor, riser rated FPLR.
> 342SP – Stranded Bare Copper, overall shielded multi-conductor, plenum rated FPLP.
> 343UP – Solid Bare Copper, unshielded pairs, plenum rated 150C NYC LL5 FPLP
> 343SP – Solid Bare Copper, overall shielded pair(s), plenum rated 150C NYC LL5 FPLP
> 344UP – Solid Bare Copper, unshielded pairs, plenum rated 200C NYC LL5 FPLP
> 344SP – Solid Bare Copper, overall shielded pair(s), plenum rated 200C NYC LL5 FPLP
> 345UCI – Solid Bare Copper, unshielded; UL 2196 Fire Resistive Cable; open wiring; FPLR-CI
> 345UCIC – Solid Bare Copper, unshielded; UL 2196 Fire Resistive Cable-In-Conduit System FPLR "CIC"
> 345SCI – Solid Bare Copper, Shielded; UL 2196 Fire Resistive Cable; open wiring; FPLR-CI
> 345SCIC – Solid Bare Copper, Shielded; UL 2196 Fire Resistive Cable-In-Conduit System FPLR "CIC"

2.5 COMMUNICATIONS CABLING

01. Cable for these sections and systems:

16700 Communications;

Cables of gage size and construction details called for on project drawings or project specifications shall be selected from the cable specifications and tables under 2.14.

> Note to Specification Writers:
> COMMUNICATIONS CABLING are typically selected from tables identified above in 2.2 01. for LOW VOLTAGE SECUIRTY, AUDIO, AND CONTROL CABLING and also these in particular:
> 333UH – Category 3 unshielded twisted pair cable, horizontal or riser rated.
> 333UP – Category 3 unshielded twisted pair cable, plenum rated.
> 335UH – Category 5e unshielded twisted pair cable, horizontal or riser rated.
> 335UP – Category 5e unshielded twisted pair cable, plenum rated.
> 336UH – Category 6 unshielded twisted pair cable, horizontal or riser rated.
> 336UP – Category 6 unshielded twisted pair cable, plenum rated.
> 356RFX - 50Ω RF Transmission Coaxial Cable used in wireless transmission.

2.6 CATV/MATV – BROADBAND AND CATV SYSTEM CABLING

01. Cable for these sections and systems:
16710 Communications Circuits – coaxial cable types;

Cables of gage sizes and construction details called for on project drawings or project specifications shall be selected from the cable specifications and tables under 2.14.

> Note to Specification Writers:
> BROADBAND AND CATV/MATV SYSTEM CABLING are selected from these tables:
> 351CAH – CATV Coaxial Cables, horizontal rated
> 351CAP – CATV Coaxial Cables, plenum rated.

2.7 CCTV – SURVEILLANCE SYSTEM CABLING

01. Cable for these sections and systems:
13700 SURVEILLANCE portion;

Cables of gage sizes and construction details called for on project drawings or project specifications shall be selected from the cable specifications and tables under 2.14.

> Note to Specification Writers:
> SURVEILLENCE SYSTEM CABLING are typically selected from these tables in particular:
> 335UH – Category 5e unshielded twisted pair cable, horizontal or riser rated.
> 335UP – Category 5e unshielded twisted pair cable, plenum rated.
> 352CCH – CCTV Coaxial Cables, horizontal rated
> 352CCP – CCTV Coaxial Cables, plenum rated
> 353CCH – CCTV Coaxial Cables with Siamese low voltage power pair, horizontal rated.
> 353CCP – CCTV Coaxial Cables with Siamese low voltage power pair, plenum rated.

2.8 SDI – Serial Digital Video; HD-SDI, precision and critical video circuits

01. Cable for these sections and systems:
16810 Video Circuits portion;

Cables of gage sizes and construction details called for on project drawings or project specifications shall be selected from the cable specifications and tables under 2.14.

Note to Specification Writers: SDI VIDEO SYSTEM CABLING are selected from these tables: 354SDR – SDI Coaxial Cables, riser rated 354SDP – SDI Coaxial Cables, plenum rated.

2.9 RGB – RGB and Component Video circuits.

Where project drawings and specifications indicate RGB or Component Video circuit cabling; cables from Cable Specification and Tables:

355RGBH -- RGB Bundled Coaxial Cables, riser rated or
355RGBP -- RGB Bundled Coaxial Cables, plenum rated

found under 2.14 shall be used accordingly.

2.10 EIA-232 (RS-232) CABLING:

Where project drawings and specifications indicate EIA-232 (RS-232) circuit cabling; cables from Cable Specification and Tables:

328BH -- EIA-232 Cables, horizontal rated or
328BP – EIA-232 Cables, plenum rated

found under 2.14 shall be used accordingly.

2.11 EIA-422 (RS-422) CABLING:

Where project drawings and specifications indicate EIA-422 (RS-422) circuit cabling; cables from Cable Specification and Tables:

328CH -- EIA-422 Cables, horizontal rated or
328CP – EIA-422 Cables, plenum rated

found under 2.14 shall be used accordingly.

2.12 EIA-485 (RS-485) CABLING:

Where project drawings and specifications indicate EIA-485 (RS-485) circuit cabling; cables from Cable Specification and Tables:

328AH – EIA-485 Cables, horizontal rated or
328AP – EIA-485 Cables, plenum rated

found under 2.14 shall be used accordingly.

2.13 IN-BUILDING WIRELESS:

50 hm Transmission Coax for Wireless Systems shall be selected from Table 356RFX.

2.14 CABLE SPECIFICATIONS AND TABLES:

Specifications for cables; and tables of Approved Manufacturers' model numbers are listed below:

PERIODIC TABLE OF PARTIAL CONTENTS – Detail for Stranded Bare Copper, UL minimum Insulation

Cabled; Stranded Bare Copper; UL Minimum Insulation **Specification 31**

#/C	Shielding	Installation	Spec / Table	Spec. Page	Table Page
2C	Unshielded	Riser	31UR-2C	22	22
		Plenum	31UP-2C	23	23
	Shielded	Riser	31SR-2C	24	24
		Plenum	31SP-2C	25	25
3C	Unshielded	Riser	31UR-3C	26	26
		Plenum	31UP-3C	27	27
	Shielded	Riser	31SR-3C	28	28
		Plenum	31SP-3C	29	29
4C	Unshielded	Riser	31UR-4C	30	30
		Plenum	31SR-4C	31	31
	Shielded	Riser	31UP-4C	32	32
		Plenum	31SP-4C	33	33
5C & Higher	Unshielded	Riser	31UR-5C	34	35
		Plenum	31UP-5C	36	37
	Shielded	Riser	31SR-5C	38	39
		Plenum	31SP-5C	40	41

Multiple Unshielded Pairs; Stranded Bare Copper; UL Minimum Insulation **Specification 32**

Shielding	Installation	Spec / Table	Spec. Page	Table Page
Unshld. Pairs / No OAS	Riser	32UR	42	46
	Plenum	32UP	43	46
Unshld. Pairs / OAS	Riser	32SR	44	47
	Plenum	32SP	45	47

Multiple Pairs; EPS/CD; Stranded Bare Copper; UL Minimum Insulation **Specification 33**

Shielding	Installation	Table	Spec. Page	Table Page
EPS / CD	Riser	33UR	48	48
	Plenum	33UP	49	49

Multiple Pairs, EPS; Stranded Bare Copper; UL Minimum Insulation **Specification 34**

Shielding	Installation	Table	Spec. Page	Table Page
EPS	Riser	34UR	50	52
	Plenum	34UP	51	52

Combo Cables; Str BC; UL Minimum Insulation; Mixed Shielding /AWG Sizes **Specification 35**

Shielding	Installation	Table	Spec. Page	Table Page
Mixed "In & Out"	Riser	35MR	53	53
	Plenum	35MP	54	54

Access Control Combo Cables; Str BC; Mixed Constr. Individual Inner Jackets **Specification 36**

Shielding	Installation	Table	Spec. Page	Table Page
Card Reader Only Shld	Riser	36MR	56 - 57	57
	Plenum	36MP	58 - 59	59
All Legs Shielded	Riser	36SR	60 - 61	61
	Plenum	36SP	62 - 63	63

Audio and Speaker Cable Stranded Bare Copper, Cabled. In-Wall Speaker **Specification 27**

Shielding	Installation	Table	Spec. Page	Table Page
Hi-Strand / Unshielded	Horizontal	27UH	64	64
Unshielded	Plenum	27UP	65	65

31UR: 300V Specification 31: Stranded Bare Copper; Cabled **Unshielded**
Str. BC / PVC or PP Insulation / Cabled / Unshielded (U) / PVC Jacket **Riser Rated (R).**

2/C Cables

Conductor: Stranded Bare Copper. Number of strands per Table 31UR-2C Column A

Insulation: PVC or Polypropylene, thickness according to UL 444 or UL 13 as applicable

Color Code: Black, Red

Assembly: Twisted together so that a 2 conductor cable is also considered 1 pair.
Ripcord under the jacket to facilitate removal.

Shielding: None.

Jacket: PVC. Thickness according to UL 444 or UL 13 as applicable.
Color: Gray unless specified.

Diameter: Nominal diameter shall not exceed that in table 31UR-2C Column B.

Flame test: Passes UL 1666, for use in risers.

Voltage: 300V but not printed on the jacket per UL 444 and UL 13 / 150V if CL2R only

Application: 22–16 AWG: Communication Cable per NEC Art. 800 and UL 444;
All sizes: Class 3 or Class 2 Remote-Control, Signaling, and Power-Limited Circuits per NEC Art. 725 and UL 13.

Installation: Suitable for open cabling use in Risers as CMR, CL3R, or CL2R; also for open cabling horizontal runs since rated higher than CM, CL3, CL2 respectively.
Not for use in plenum spaces unless running in conduit.

Temp.: 0C to 60C minimum range, continuous operating ambient temperature.

QA: UL Listed as cable type CMR (22-16AWG) or CL3R, or both; rating(s) printed on the jacket.
UL file number will be printed on the jacket alongside the gauge size and cable rating.
UL label shall be on every cable package near manufacturer's product identification label.

Packaging: 1000ft or 500ft: spools, reels or boxes.

Appr. Manf: Manufacturer make and model shall be as listed in table 31UR-2C.

Availability: Cable shall be available from local distribution known to project management.

Table 31UR-2C **Str. BC; PVC or PP Insul.; Cabled; Unshielded; Riser Rated**

Const	SDN	A	B	C	TAPPAN	GENERAL CABLE	BELDEN-WEST PENN	BELDEN
22/2	31UR.222	7	.130	10	R20004	E1002S	221	5500UE
20/2	31UR.202	7	.142	13	R30017	E1022S	222	5400UE
18/2	31UR.182	7	.166	18	R40018	E1032S	224	5300UE
16/2	31UR.162	19	.190	26	R50003	E1042S	225	5200UE
14/2	31UR.142	19	.234	40	R60012	E1052S	226	5100UE
12/2	31UR.122	19	.276	58	R70003	E1062S	227	5000UE

Manufacturer's makes vary. Diameters (Col. B) and weights (Col. C) are conservative and for engineering purposes only. Consult manufacturer's data for individual make & model information.

Column A: Stranded bare copper minimum number of strands.
Column B: Nominal Cable Diameter in inches, for conduit fill calculations.
Column C: Nominal cable weight in pounds per 1000ft.

31UP: 300V Specification 31: Stranded Bare Copper; Cabled
Str. BC / LS-PVC / Cabled / Unshielded (U) / LS-PVC Jacket

Unshielded
Plenum Rated (P).

2/C Cables

Conductor: Stranded Bare Copper per Table 31P-2C Column A

Insulation: LS-PVC, thickness according to UL 444 or UL 13 as applicable.

Color Code: Black, Red

Assembly: Twisted together so that a . 2 conductor cable is also considered 1 pair
Ripcord under the jacket to facilitate removal.

Shielding: None.

Jacket: LS-PVC. Thickness according to UL 444 or UL 13 as applicable.
Color: Natural or White, or gray; unless specified.

Diameter: Nominal diameter shall not exceed that in table 31UP-2C Column B.

Flame test: Passes NFPA 262 (UL 910 Steiner Tunnel) flame test for use in plenum spaces.

Voltage: 300V but not printed on the jacket per UL 444 and UL 13. / 150V if CL2P only

Application: 22–16 AWG: Communication Cable per NEC Art. 800 and UL 444;
All sizes: Class 3 or Class 2 Remote-Control, Signaling, and Power-Limited Circuits per NEC Art. 725 and UL 13.

Installation: Suitable for open cabling use in plenum spaces as CMP, CL3P or CL2P; also for open cabling use in risers and in horizontal runs since rated higher than those requirements.

Temp.: 0C to 60C minimum range, continuous operating ambient temperature.

QA: UL Listed as cable type CMP, CL3P, or both. Cable rating(s) shall be printed on the jacket.
UL file number will be printed on the jacket alongside the gauge size and cable rating.
UL label shall be on every cable package near manufacturer's product identification label.

Packaging: 1000ft or 500ft: spools, reels or boxes.

Appr. Manf: Manufacturer make and model shall be as listed in table 31UP-2C.

Availability: Cable shall be available from local distribution known to project management.

Table 31UP-2C Str. BC; LS-PVC Insul.; Cabled, Unshielded; Plenum Rated

Const	SDN	A	B	C	TAPPAN	GENERAL CABLE	BELDEN-WEST PENN	BELDEN
22/2	31UP.222	7	.120	9	P20031	E3002S	25221B	6500UE
20/2	31UP.202	7	.138	12	P30016	E3022S	25222B	6400UE
18/2	31UP.182	7	.156	18	P40020	E3032S	25224B	6300UE
16/2	31UP.162	19	.178	26	P50032	E3042S	25225B	6200UE
14/2	31UP.142	19	.216	38	P60040	E3052S	25226B	6100UE
12/2	31UP.122	19	.264	56	P70031	E3062S	25227B	6000UE

Manufacturer's makes vary. Diameters (Col. B) and weights (Col. C) are conservative and for engineering purposes only. Consult manufacturer's data for individual make & model information.

Column A: Stranded bare copper conductor number of strands.
Column B: Nominal Cable Diameter in inches, for conduit fill calculations.
Column C: Nominal cable weight in pounds per 1000ft.

31SR: 300V Specification 31: Stranded Bare Copper; Cabled Shielded
Str. BC / PVC or PP Insulation / Cabled / Shielded (S) / PVC Jacket. Riser Rated (R).

2/C Cables

Conductor: Stranded Bare Copper per Table 31SR-2C Column A

Insulation: PVC or Polypropylene, thickness according to UL 444 or UL 13 as applicable.

Color Code: Black, Red

Assembly: Twisted together so that a 2 conductor cable is also considered 1 pair.
Ripcord under the jacket to facilitate removal.

Shielding: Aluminum-polyester shielding tape, Tinned Copper Drain Wire. Aluminum foil surface facing out; drain wire is outside the tape and in continuous contact with the foil.

Jacket: PVC, thickness according to UL 444 or UL 13 as applicable.
Color: Gray unless specified.

Diameter: Nominal diameter shall not exceed that in table 31SR-2C Column B.

Flame test: Passes UL 1666, for use in risers.

Voltage: 300V but not printed on the jacket per UL 444 and UL 13. / 150V if CL2R only

Application: 22–16 AWG: Communication Cable per NEC Art. 800 and UL 444;
All sizes: Class 3 or Class 2 Remote-Control, Signaling, and Power-Limited Circuits per NEC Art. 725 and UL 13.

Installation: Suitable for open cabling use in Risers as CMR, CL3R, or CL2R; also for open wiring horizontal runs since rated higher than CM, CL3, CL2 respectively.
Not for use in plenum spaces unless running in conduit.

Temp.: 0C to 60C minimum range, continuous operating ambient temperature.

QA: UL Listed as cable type CMR, CL3R, or both. Cable rating(s) shall be printed on the jacket.
UL file number will be printed on the jacket alongside the gauge size and cable rating.
UL label shall be displayed on every cable package near manufacturer's product label.

Packaging: 1000ft or 500ft: spools, reels or boxes.

Appr. Manf: Manufacturer make and model shall be as listed in table 31SR-2C.

Availability: Cable shall be available from local distribution known to project management.

Table 31SR-2C Str. BC; PVC or PP Insul.; Cabled; Shielded, Riser Rated

Const	SDN	A	B	C	TAPPAN	GENERAL CABLE	BELDEN-WEST PENN	BELDEN
22/2	31SR.222	7	.132	12	R20007	E2002S	291	5500FE
20/2	31SR.202	7	.143	16	R30024	E2022S	292	5400FE
18/2	31SR.182	7	.166	20	R40013	E2032S	293	5300FE
16/2	31SR.162	19	.191	32	A50039	E2042S	294	5200FE
14/2	31SR.142	19	.245	40	R60052	E2052S	295	5100FE
12/2	31SR.122	19	.281	59	R70010	E2062S	296	5000FE

Manufacturer's makes vary. Diameters (Col. B) and weights (Col. C) are conservative and for engineering purposes only. Consult manufacturer's data for individual make & model information.

Column A: Stranded bare copper conductor number of strands.
Column B: Nominal Cable Diameter in inches, for conduit fill calculations.
Column C: Nominal cable weight in pounds per 1000ft.

31SP: 300V Specification 31: Stranded Bare Copper; Cabled — Shielded
Str. BC / LS-PVC Insulation / Cabled / Shielded (S) / LS-PVC Jacket. — Plenum Rated (P).

2/C Cables

Conductor: Stranded Bare Copper per Table 31SP-2C Column A

Insulation: LS-PVC, thickness according to UL 444 or UL 13 as applicable.

Color Code: Black, Red

Assembly: Twisted together so that a 2 conductor cable is also considered 1 pair. Ripcord under the jacket to facilitate removal.

Shielding: Aluminum-polyester shielding tape, Tinned Copper Drain Wire. Aluminum foil surface facing out; drain wire is outside the tape and in continuous contact with the foil.

Jacket: LS-PVC, thickness according to UL 444 or UL 13 as applicable. Color: Natural, or white, or gray unless specified.

Diameter: Nominal diameter shall not exceed that in table 31SP-2C Column B.

Flame test: Passes NFPA 262 (UL 910 Steiner Tunnel) flame test for use in plenum spaces.

Voltage: 300V but not printed on the jacket per UL 444 and UL 13. / 150V if CL2P only

Application: 22–16 AWG: Communication Cable per NEC Art. 800 and UL 444; all sizes: Class 3 or Class 2 Remote-Control, Signaling, and Power-Limited Circuits per NEC Art. 725 and UL 13.

Installation: Suitable for open cabling use in plenum spaces as CMP, CL3P or CL2P; also for open cabling use in risers and in horizontal runs since rated higher than those requirements.

Temp.: 0C to 60C minimum range, continuous operating ambient temperature.

QA: UL Listed as cable type CMP, CL3P, or both. Cable rating(s) shall be printed on the jacket. UL file number will be printed on the jacket alongside gage size and cable ratings. UL label shall be displayed on every cable package near manufacturer's label.

Packaging: 1000ft or 500ft: spools, reels or boxes.

Appr. Manf: Manufacturer make and model shall be as listed in table 31SP-2C.

Availability: Cable shall be available from local distribution known to project management.

Table 31SP-2C Str. BC; LS-PVC Insul.; Cabled; Shielded; Plenum Rated

Const	SDN	A	B	C	TAPPAN	GENERAL CABLE	BELDEN-WEST PENN	BELDEN
22/2	31SP.222	7	.136	13	P20032	E2102S	25291B	6500FE
20/2	31SP.202	7	.143	13	P30006	E2122S	25292B	6400FE
18/2	31SP.182	7	.170	22	P40133	E2202S	25293B	6300FE
16/2	31SP.162	19	.200	34	P50002	E2242S	25294B	6200FE
14/2	31SP.142	19	.224	48	P60022	E2252S	25295B	6100FE
12/2	31SP.122	19	.264	58	P70006	E2262S	25296B	6000FE

Manufacturer's makes vary slightly. Diameters (Col. B) and weights (Col. C) shown are the largest from among these manufacturers' published data.

Column A: Stranded bare copper conductor number of strands.
Column B: Nominal Cable Diameter in inches, for conduit fill calculations.
Column C: Nominal cable weight in pounds per 1000ft.

31UR: 300V Specification 31: Stranded Bare Copper; Cabled **Unshielded**
Str. BC / PVC or PP Insulation / Cabled / Unshielded (U) / PVC Jacket. **Riser Rated (R).**

3/C Cables

Conductor: Stranded Bare Copper. Number of strands per Table Column A

Insulation: PVC or Polypropylene, thickness according to UL 444 or UL 13 as applicable

Color Code: Black, Red, White

Assembly: Twisted together so that a 3 conductor cable is also considered 1 triad.
Ripcord under the jacket to facilitate removal.

Shielding: None.

Jacket: PVC. Thickness according to UL 444 or UL 13 as applicable.
Color: Gray unless specified.

Diameter: Nominal diameter shall not exceed that in table 31UR-3C Column B.

Flame test: Passes UL 1666, for use in risers.

Voltage: 300V but not printed on the jacket per UL 444 and UL 13. / 150V if CL2R only

Application: 22–16 AWG: Communication Cable per NEC Art. 800 and UL 444; all sizes: Class 3 or Class 2 Remote-Control, Signaling, and Power-Limited Circuits per NEC Art. 725 and UL 13.

Installation: Suitable for open cabling use in Risers as CMR, CL3R, or CL2R; also for open cabling in horizontal runs since rated higher than CM, CL3, CL2 respectively.
Not for use in plenum spaces unless running in conduit.

Temp.: 0C to 60C minimum range, continuous operating ambient temperature.

QA: UL Listed as cable type CMR (22-16AWG), CL3R, or both; rating(s) printed on the jacket.
UL file number will be printed on the jacket alongside the gauge size and cable rating.
UL label shall be displayed on every cable package near manufacturer's product label.

Packaging: 1000ft or 500ft: spools, reels or boxes.

Appr. Manf: Manufacturer make and model shall be as listed in table 31UR-3C.

Availability: Cable shall be available from local distribution known to project management.

Table 31UR-3C Str. BC; PVC or PP Insul.; Cabled; Unshielded; Riser Rated

Const	SDN	A	B	C	TAPPAN	GENERAL CABLE	BELDEN-WEST PENN	BELDEN
22/3	31UR.223	7	.141	13	R20044	E1003S	231	5501UE
20/3	31UR.203	7	.158	18	R30008	E1023S	232	5401UE
18/3	31UR.183	7	.174	25	R40048	E1033S	234	5301UE
16/3	31UR.163	19	.203	36	R50022	E1043S	235	5201UE
14/3	31UR.143	19	.249	56	R60013	-----	236	5101UE
12/3	31UR.123	19	.307	82	R70004	-----	238	5001UE

Manufacturer's makes vary. Diameters (Col. B) and weights (Col. C) are conservative and for engineering purposes only. Consult manufacturer's data for individual make & model information

Column A: Stranded bare copper conductor number of strands.
Column B: Nominal Cable Diameter in inches, for conduit fill calculations.
Column C: Nominal cable weight in pounds per 1000ft.

31UP: 300V Specification 31: Stranded Bare Copper; Cabled Unshielded
Str. BC / LS-PVC / Cabled / Unshielded (U) / LS-PVC Jacket. Plenum Rated (P).

3/C Cables

Conductor: Stranded Bare Copper per Table 31UP-3C Column A

Insulation: LS-PVC, thickness according to UL 444 or UL 13 as applicable.

Color Code: Black, White, Red

Assembly: Twisted together so that a . 3 conductor cable is also considered 1 triad.
Ripcord under the jacket to facilitate removal.

Shielding: None.

Jacket: LS-PVC. Thickness according to UL 444 or UL 13 as applicable.
Color: Natural or White, or gray; unless specified.

Diameter: Nominal diameter shall not exceed that in table 31UP-3C Column B.

Flame test: Passes NFPA 262 (UL 910 Steiner Tunnel) flame test for use in plenum spaces.

Voltage: 300V but not printed on the jacket per UL 444 and UL 13. / 150V if CL2P only

Application: 22–16 AWG: Communication Cable per NEC Art. 800 and UL 444;
All sizes: Class 3 or Class 2 Remote-Control, Signaling, and Power-Limited Circuits per NEC Art. 725 and UL 13.

Installation: Suitable for open cabling use in plenum spaces as CMP, CL3P or CL2P; also for open cabling use in risers and in horizontal runs since rated higher than those requirements.

Temp.: 0C to 60C minimum range, continuous operating ambient temperature.

QA: UL Listed as cable type CMP, CL3P, or both. Cable rating(s) shall be printed on the jacket.
UL file number will be printed on the jacket alongside the gauge size and cable rating.
UL label shall be displayed on every cable package near manufacturer's product label.

Packaging: 1000ft or 500ft: spools, reels or boxes.

Appr. Manf: Manufacturer make and model shall be as listed in table 31UP-3C

Availability: Cable shall be available from local distribution known to project management

Table 31UP-3C Str. BC; LS-PVC Insul.; Cabled, Unshielded; Plenum Rated

Const	SDN	A	B	C	TAPPAN	GENERAL CABLE	BELDEN-WEST PENN	BELDEN
22/3	31UP.223	7	.136	14	P20134	E3003S	25231B	6501UE
20/3	31UP.203	7	.150	18	P30033	E3023S	25232B	6401UE
18/3	31UP.183	7	.168	25	P40024	E3033S	25234B	6301UE
16/3	31UP.163	19	.191	36	P50012	E3043S	-----	6201UE
14/3	31UP.143	19	.240	56	P60012	-----	25236B	6101UE
12/3	31UP.123	19	.289	83	P70018	-----	25238B	6001UE

Manufacturer's makes vary. Diameters (Col. B) and weights (Col. C) are conservative and for engineering purposes only. Consult manufacturer's data for individual make & model information

Column A: Stranded bare copper conductor number of strands.
Column B: Nominal Cable Diameter in inches, for conduit fill calculations.
Column C: Nominal cable weight in pounds per 1000ft.

31SR: 300V Specification 31: Stranded Bare Copper; Cabled **Shielded**
Str. BC / PVC or PP Insulation / Cabled / Shielded (S) / PVC Jacket. **Riser Rated (R).**

3/C Cables

Conductor: Stranded Bare Copper. Number of strands per Table 31SR-3C Column A

Insulation: PVC or Polypropylene, thickness according to UL 444 or UL 13 as applicable

Color Code: Black, Red, White

Assembly: Twisted together so that a 3 conductor cable is also considered 1 triad.
Ripcord under the jacket to facilitate removal.

Shielding: Aluminum-polyester shielding tape, Tinned Copper Drain Wire. Aluminum foil surface facing out; drain wire is outside the tape and in continuous contact with the foil.

Jacket: PVC. Thickness according to UL 444 or UL 13 as applicable.
Color: Gray unless specified.

Diameter: Nominal diameter shall not exceed that in table 31SR-3C Column B.

Flame test: Passes UL 1666, for use in risers.

Voltage: 300V but not printed on the jacket per UL 444 and UL 13. / 150V if CL2R only

Application: 22–16 AWG: Communication Cable per NEC Art. 800 and UL 444;
All sizes: Class 3 or Class 2 Remote-Control, Signaling, and Power-Limited Circuits per NEC Art. 725 and UL 13.

Installation: Suitable for open cabling use in Risers as CMR, CL3R, or CL2R; also for open cabling in horizontal runs since rated higher than CM, CL3, CL2 respectively.
Not for use in plenum spaces unless running in conduit.

Temp.: 0C to 60C minimum range, continuous operating ambient temperature.

QA: UL Listed as cable type CMR (22-16AWG), CL3R, or both; rating(s) printed on the jacket.
UL file number will be printed on the jacket alongside the gauge size and cable rating.
UL label shall be displayed on every cable package near manufacturer's product label.

Packaging: 1000ft or 500ft: spools, reels or boxes.
Appr. Manf: Manufacturer make and model shall be as listed in table 31SR-3C
Availability: Cable shall be available from local distribution known to project management

Table 31SR-3C **Str. BC; PVC or PP Insul.; Cabled; Shielded, Riser Rated**

Const	SDN	A	B	C	TAPPAN	GENERAL CABLE	BELDEN-WEST PENN	BELDEN
22/3	31SR.223	7	.145	15	R20016	E2003S	301	5501FE
20/3	31SR.203	7	.162	21	R30041	E2023S	302	5401FE
18/3	31SR.183	7	.178	27	R40003	E2033S	303	5301FE
16/3	31SR.163	19	.206	42	R50044	E2043S	304	5201FE
14/3	31SR.143	19	.270	66	R60023	------	305	5101FE
12/3	31SR.123	19	.296	85	R70011		------	------

Manufacturer's makes vary. Diameters (Col. B) and weights (Col. C) are conservative and for engineering purposes only. Consult manufacturer's data for individual make & model information

Column A: Stranded bare copper conductor number of strands.
Column B: Nominal Cable Diameter in inches, for conduit fill calculations.
Column C: Nominal cable weight in pounds per 1000ft.

31SP: 300V Specification 31: Stranded Bare Copper; Cabled **Shielded**
Str. BC / LS-PVC Insulation / Cabled / Shielded (S) / LS-PVC Jacket **Plenum Rated (P).**

3/C Cables

Conductor: Stranded Bare Copper per Table 31SP-3C Column A

Insulation: LS-PVC, thickness according to UL 444 or UL 13 as applicable.

Color Code: Black, White, Red

Assembly: Twisted together so that a . 3 conductor cable is also considered 1 triad.
Ripcord under the jacket to facilitate removal.

Shielding: Aluminum-polyester shielding tape, Tinned Copper Drain Wire. Aluminum foil surface facing out; drain wire is outside the tape and in continuous contact with the foil.

Jacket: LS-PVC. Thickness according to UL 444 or UL 13 as applicable.
Color: Natural or White, or gray; unless specified.

Diameter: Nominal diameter shall not exceed that in table 31SP-3C Column B.

Flame test: Passes NFPA 262 (UL 910 Steiner Tunnel) flame test for use in plenum spaces.

Voltage: 300V but not printed on the jacket per UL 444 and UL 13. / 150V if CL2P only

Application: 22–16 AWG: Communication Cable per NEC Art. 800 and UL 444;
All sizes: Class 3 or Class 2 Remote-Control, Signaling, and Power-Limited Circuits per NEC Art. 725 and UL 13.

Installation: Suitable for open cabling use in plenum spaces as CMP, CL3P or CL2P; also for open cabling use in risers and in horizontal runs since rated higher than those requirements.

Temp.: 0C to 60C minimum range, continuous operating ambient temperature.

QA: UL Listed as cable type CMP, CL3P, or both. Cable rating(s) shall be printed on the jacket.
UL file number will be printed on the jacket alongside the gauge size and cable rating.
UL label shall be displayed on every cable package near manufacturer's product label.

Packaging: 1000ft or 500ft: spools, reels or boxes.
Appr. Manf: Manufacturer make and model shall be as listed in table 31SP-3C.
Availability: Cable shall be available from local distribution known to project management

Table 31SP-3C Str. BC; LS-PVC Insul.; Cabled; Shielded; Plenum Rated

Const	SDN	A	B	C	TAPPAN	GENERAL CABLE	BELDEN-WEST PENN	BELDEN
22/3	31SP.223	7	.141	16	P20043	E2103S	25310B	6501FE
20/3	31SP.203	7	.154	22	P30045	E2123S	25302B	6401FE
18/3	31SP.183	7	.186	31	P40032	E2203S	25303B	6301FE
16/3	31SP.163	19	.195	45	P50022	E2243S	-----	6201FE
14/3	31SP.143	19	.234	66	P60023	-----	-----	6101FE
12/3	31SP.123	19	.284	84	P70003	-----	-----	-----

Manufacturer's makes vary. Diameters (Col. B) and weights (Col. C) are conservative and for engineering purposes only. Consult manufacturer's data for individual make & model information.

Column A: Stranded bare copper conductor number of strands.
Column B: Nominal Cable Diameter in inches, for conduit fill calculations.
Column C: Nominal cable weight in pounds per 1000ft.

31UR: 300V Specification 31: Stranded Bare Copper; Cabled **Unshielded**
Str. BC / PVC or PP Insulation / Cabled / Unshielded (U) / PVC Jacket. **Riser Rated (R).**

4/C Cables

Conductor: Stranded Bare Copper per Table 31UR-4C Column A

Insulation: PVC, thickness according to UL 444 or UL 13 as applicable.

Color Code: Black, Red, White, Green

Assembly: Twisted together so that a 4 conductor cable qualifies as 1 quad.
Ripcord under the jacket to facilitate removal.

Shielding: None.

Jacket: PVC. Thickness according to UL 444 or UL 13 as applicable.
Color: Gray unless specified.

Diameter: Nominal diameter shall not exceed that in table 31UR-4C Column B.

Flame test: Passes UL 1666, for use in risers.

Voltage: 300V but not printed on the jacket per UL 444 and UL 13. / 150V if CL2R only

Application: 22–16 AWG: Communication Cable per NEC Art. 800 and UL 444;
All sizes: Class 3 or Class 2 Remote-Control, Signaling, and Power-Limited Circuits per NEC Art. 725 and UL 13.

Installation: Suitable for open cabling use in Risers as CMR, CL3R, or CL2R; also for open cabling horizontal runs since rated higher than CM, CL3, CL2 respectively.
Not for use in plenum spaces unless running in conduit.

Temp.: 0C to 60C minimum range, continuous operating ambient temperature.

QA: UL Listed as cable type CMR (22-16AWG) or CL3R, or both; rating(s) printed on the jacket.
UL file number will be printed on the jacket alongside the gauge size and cable rating.
UL label shall be displayed on every cable package near manufacturer's product label.

Packaging: 1000ft or 500ft: spools, reels or boxes.

Appr. Manf: Manufacturer make and model shall be as listed in table 31UR-4C.

Availability: Cable shall be available from local distribution known to project management

Table 31UR-4C Str. BC; PVC or PP Insul.; Cabled; Unshielded; Riser Rated

Const	SDN	A	B	C	TAPPAN	GENERAL CABLE	BELDEN-WEST PENN	BELDEN
22/4	31UR.224	7	.153	16	R20045	E1004S	241	5502UE
20/4	31UR.204	7	.173	23	R30006	E1024S	242	5402UE
18/4	31UR.184	7	.192	32	R40049	E1034S	244	5302UE
16/4	31UR.164	19	.222	47	R50045	E1044S	245	5202UE
14/4	31UR.144	19	.292	71	R60014	E1054S	246	5102UE
12/4	31UR.124	19	.338	106	R70005	E1064S	248	-----

Manufacturer's makes vary. Diameters (Col. B) and weights (Col. C) are conservative and for engineering purposes only. Consult manufacturer's data for individual make & model information.

Column A: Stranded bare copper conductor number of strands.
Column B: Nominal Cable Diameter in inches, for conduit fill calculations.
Column C: Nominal cable weight in pounds per 1000ft.

31UP: 300V Specification 31: Stranded Bare Copper; Cabled **Unshielded**
Str. BC / LS-PVC / Cabled / Unshielded (U) / LS-PVC Jacket **Plenum Rated (P).**

4/C Cables

Conductor: Stranded Bare Copper per Table 31UP-4C Column A

Insulation: LS-PVC, thickness according to UL 444 or UL 13 as applicable.

Color Code: Black, White, Red, Green

Assembly: Twisted together so that a . 4 conductor cable is also considered 1 quad.
Ripcord under the jacket to facilitate removal.

Shielding: None.

Jacket: LS-PVC. Thickness according to UL 444 or UL 13 as applicable.
Color: Natural or White, or gray; unless specified.

Diameter: Nominal diameter shall not exceed that in table 31UP-4C Column B.

Flame test: Passes NFPA 262 (UL 910 Steiner Tunnel) flame test for use in plenum spaces.

Voltage: 300V but not printed on the jacket per UL 444 and UL 13. / 150V if CL2P only.

Application: 22–16 AWG: Communication Cable per NEC Art. 800 and UL 444;
All sizes: Class 3 or Class 2 Remote-Control, Signaling, and Power-Limited Circuits per NEC Art. 725 and UL 13.

Installation: Suitable for open cabling use in plenum spaces as CMP, CL3P or CL2P; also for open cabling use in risers and in horizontal runs since rated higher than those requirements.

Temp.: 0C to 60C minimum range, continuous operating ambient temperature.

QA: UL Listed as cable type CMP, CL3P, or both. Cable rating(s) shall be printed on the jacket.
UL file number will be printed on the jacket alongside the gauge size and cable rating.
UL label shall be displayed on every cable package near manufacturer's product label.

Packaging: 1000ft or 500ft: spools, reels or boxes.

Appr. Manf: Manufacturer make and model shall be as listed in table 31UP-4C

Availability: Cable shall be available from local distribution known to project management

Table 31UP-4C Str. BC; LS-PVC Insul.; Cabled, Unshielded; Plenum Rated

Const	SDN	A	B	C	TAPPAN	GENERAL CABLE	BELDEN-WEST PENN	BELDEN
22/4	31UP.224	7	.148	17	P20035	E3004S	25241B	6502UE
20/4	31UP.204	7	.170	24	P30047	E3024S	25242B	6402UE
18/4	31UP.184	7	.189	39	P40031	E3034S	25244B	6302UE
16/4	31UP.164	19	.221	52	P50059	E3044S	25245B	6202UE
14/4	31UP.144	19	.265	73	P60013	E3054S	25246B	6102UE
12/4	31UP.124	19	.314	111	P70019	E3064S	25248B	6002UE

Manufacturer's makes vary. Diameters (Col. B) and weights (Col. C) are conservative and for engineering purposes only. Consult manufacturer's data for individual make & model information.

Column A: Stranded bare copper conductor number of strands.
Column B: Nominal Cable Diameter in inches, for conduit fill calculations.
Column C: Nominal cable weight in pounds per 1000ft.

31SR: 300V Specification 31: Stranded Bare Copper; Cabled **Shielded**
Str. BC / PVC or PP Insulation / Cabled / Shielded (S) / PVC Jacket. **Riser Rated (R).**

4/C Cables

Conductor: Stranded Bare Copper. Number of strands per Table 31SR-4C Column A

Insulation: PVC or Polypropylene, thickness according to UL 444 or UL 13 as applicable

Color Code: Black, Red, White, Green

Assembly: Twisted together so that a 4 conductor cable is also considered 1 quad.
Ripcord under the jacket to facilitate removal.

Shielding: Aluminum-mylar shielding tape, Tinned Copper Drain Wire. Aluminum foil surface facing out; drain wire is outside the tape and in continuous contact with the foil.

Jacket: PVC. Thickness according to UL 444 or UL 13 as applicable.
Color: Gray unless specified.

Diameter: Nominal diameter shall not exceed that in table 31SR-4C Column B.

Flame test: Passes UL 1666, for use in risers.

Voltage: 300V but not printed on the jacket per UL 444 and UL 13. / 150V if CL2R only

Application: 22–16 AWG: Communication Cable per NEC Art. 800 and UL 444; all sizes: Class 3 or Class 2 Remote-Control, Signaling, and Power-Limited Circuits per NEC Art. 725 and UL 13.

Installation: Suitable for open cabling use in Risers as CMR, CL3R, or CL2R; also for open cabling horizontal runs since rated higher than CM, CL3, CL2 respectively.
Not for use in plenum spaces unless running in conduit.

Temp.: 0C to 60C minimum range, continuous operating ambient temperature.

QA: UL Listed as cable type CMR (22-16AWG), CL3R, or both; rating(s) printed on the jacket.
UL file number will be printed on the jacket alongside the gauge size and cable rating.
UL label shall be displayed on every cable package near manufacturer's product label.

Packaging: 1000ft or 500ft: spools, reels or boxes.
Appr. Manf: Manufacturer make and model shall be as listed in table 31SR-4C
Availability: Cable shall be available from local distribution known to project management

Table 31SR-4C Str. BC; PVC or PP Insul.; Cabled; Shielded, Riser Rated

Const	SDN	A	B	C	TAPPAN	GENERAL CABLE	BELDEN-WEST PENN	BELDEN
22/4	31SR.224	7	.157	18	R20030	E2004S	3241	5502FE
20/4	31SR.204	7	.176	26	R30042	E2024S	3011	5402FE
18/4	31SR.184	7	.196	37	R40005	E2034S	3244	5302FE
16/4	31SR.164	19	.235	54	R50031	E2044S	3245	5202FE
14/4	31SR.144	19	.269	71	R60024	E2054S	-----	-----
12/4	31SR.124	19	.326	110	R70012	E2064S	-----	-----

Manufacturer's makes vary. Diameters (Col. B) and weights (Col. C) are conservative and for engineering purposes only. Consult manufacturer's data for individual make & model information.

Column A: Stranded bare copper conductor number of strands.
Column B: Nominal Cable Diameter in inches, for conduit fill calculations.
Column C: Nominal cable weight in pounds per 1000ft.

31SP: 300V Specification 31: Stranded Bare Copper; Cabled Shielded
Str. BC / LS-PVC Insulation / Cabled / Shielded (S) / LS-PVC Jacket. Plenum Rated (P).

4/C Cables

Conductor: Stranded Bare Copper per Table 31SP-4C Column A

Insulation: LS-PVC, thickness according to UL 444 or UL 13 as applicable.

Color Code: Black, White, Red, Green

Assembly: Twisted together so that a . 4 conductor cable is also considered 1 quad.
Ripcord under the jacket to facilitate removal.

Shielding: Aluminum-polyester shielding tape, Tinned Copper Drain Wire. Aluminum foil surface facing out; drain wire is outside the tape and in continuous contact with the foil.

Jacket: LS-PVC. Thickness according to UL 444 or UL 13 as applicable.
Color: Natural or White; unless specified.

Diameter: Nominal diameter shall not exceed that in table 31SP-4C Column B.

Flame test: Passes NFPA 262 (UL 910 Steiner Tunnel) flame test for use in plenum spaces.

Voltage: 300V but not printed on the jacket per UL 444 and UL 13. / 150V if CL2P only

Application: 22–16 AWG: Communication Cable per NEC Art. 800 and UL 444;
All sizes: Class 3 or Class 2 Remote-Control, Signaling, and Power-Limited Circuits per NEC Art. 725 and UL 13.

Installation: Suitable for open cabling use in plenum spaces as CMP, CL3P or CL2P; also for open cabling use in risers and in horizontal runs since rated higher than those requirements.

Temp.: 0C to 60C minimum range, continuous operating ambient temperature.

QA: UL Listed as cable type CMP, CL3P, or both. Cable rating(s) shall be printed on the jacket.
UL file number will be printed on the jacket alongside the gauge size and cable rating.
UL label shall be displayed on every cable package near manufacturer's product label.

Packaging: 1000ft or 500ft: spools, reels or boxes.

Appr. Manf: Manufacturer make and model shall be as listed in table 31SP-4C.

Availability: Cable shall be available from local distribution known to project management

Table 31SP-4C Str. BC; LS-PVC Insul.; Cabled; Shielded; Plenum Rated

Const	SDN	A	B	C	TAPPAN	GENERAL CABLE	BELDEN-WEST PENN	BELDEN
22/4	31SP.224	7	.157	20	P20270	E2104S	253241B	6502FE
20/4	31SP.204	7	.166	27	P30003	E2124S	-----	6402FE
18/4	31SP.184	7	.198	37	P40073	E2204S	253244B	6302FE
16/4	31SP.164	19	.215	59	P50023	E2244S	-----	6202FE
14/4	31SP.144	19	.260	71	P60024	E2254S	-----	-----
12/4	31SP.124	19	.314	110	P70008	E2264S	-----	-----

Manufacturer's makes vary. Diameters (Col. B) and weights (Col. C) are conservative and for engineering purposes only. Consult manufacturer's data for individual make & model information.

Column A: Stranded bare copper conductor number of strands.
Column B: Nominal Cable Diameter in inches, for conduit fill calculations.
Column C: Nominal cable weight in pounds per 1000ft.

31UR: 300V Specification 31: Stranded Bare Copper; Cabled **Unshielded**
Str. BC / PVC or PP Insulation / Cabled / Unshielded (U) / PVC Jacket. **Riser Rated (R).**

5/C and Higher Cables

Conductor: Stranded Bare Copper per Table 31UR-5C Column A

Insulation: PVC or Polypropylene, thickness according to UL 444 or UL 13 as applicable

Color Code: Black, Red, White, Green, Brown, Blue, Orange, Yellow, Violet, Gray, Pink, Beige, White striped Black, White striped Red, White striped Green, White striped Orange, White striped Blue, White striped Brown, White striped Yellow, White striped Violet.

Assembly: Multiple conductors twisted together for flexibility. Ripcord under the jacket to facilitate removal.

Shielding: None.

Jacket: PVC. Thickness according to UL 444 or UL 13 as applicable. Color: Gray unless specified.

Diameter: Nominal diameter shall not exceed that in table 31UR-5C Column B.

Flame test: Passes UL 1666, for use in risers.

Voltage: 300V but not printed on the jacket per UL 444 and UL 13. / 150V if CL2R only

Application: 22–16 AWG: Communication Cable per NEC Art. 800 and UL 444; All sizes: Class 3 or Class 2 Remote-Control, Signaling, and Power-Limited Circuits per NEC Art. 725 and UL 13.

Installation: Suitable for open cabling use in Risers as CMR, CL3R, or CL2R; also for open cabling horizontal runs since rated higher than CM, CL3, CL2 respectively. Not for use in plenum spaces unless running in conduit.

Temp.: 0C to 60C minimum range, continuous operating ambient temperature.

QA: UL Listed as cable type CMR (22-16AWG), CL3R, or both; rating(s) printed on the jacket. UL file number will be printed on the jacket alongside the gauge size and cable rating. UL label shall be displayed on every cable package near manufacturer's product label.

Packaging: 1000ft or 500ft: spools or reels. Boxes optional where practicable.

Appr. Manf: Manufacturer make and model shall be as listed in table 31UR-5C.

Availability: Cable shall be available from local distribution known to project management

Table 31UR-5C

Table 31UR-5C-22AWG Str. BC; PVC or PP Insul.; Cabled; Unshielded; Riser Rated

Const	SDN	A	B	C	TAPPAN	GENERAL CABLE	BELDEN-WEST PENN	BELDEN
22/5	31UR.225	7	.168	23	R20046	-----	855	5503UE
22/6	31UR.226	7	.191	28	R20047	E1006S	270	5504UE
22/7	31UR.227	7	.196	29	R20048	-----	-----	-----
22/8	31UR.228	7	.198	30	R20025	E1008S	271	5506UE
22/9	31UR.229	7	.200	32	R20050	-----	-----	-----
22/10	31UR.2210	7	.243	41	R20051	E1010S	272	5508UE
22/12	31UR.2212	7	.251	47	R20487	E1012S	273	5509UE
22/15	31UR.2215	7	.278	58	R20056	-----	274	-----
22/20	31UR.2220	7	.308	65	R20061	-----	275	-----

Table 31UR-5C-20AWG Str. BC; PVC or PP Insul.; Cabled; Unshielded; Riser Rated

Const	SDN	A	B	C	TAPPAN	GENERAL CABLE	BELDEN-WEST PENN	BELDEN
20/5	31UR.205	7	.189	28	R30029	-----	262	5403UE
20/6	31UR.206	7	.195	28	R30030	-----	-----	-----
20/7	31UR.207	7	.207	40	R30031	-----	263	5405UE
20/8	31UR.208	7	.215	43	R30032	-----	-----	5406UE
20/9	31UR.209	7	.253	48	R30267	-----	264	5407UE
20/10	31UR.2010	7	.254	53	R30034	-----	-----	5408UE
20/12	31UR.2012	7	.284	62	R30035	-----	265	5409UE
20/15	31UR.2015	7	.290	69	R30036	-----	-----	-----
20/20	31UR.2020	7	.347	110	R30037	-----	-----	540BUE

Table 31UR-5C-18AWG Str. BC; PVC or PP Insul.; Cabled; Unshielded; Riser Rated

Const	SDN	A	B	C	TAPPAN	GENERAL CABLE	BELDEN-WEST PENN	BELDEN
18/5	31UR.185	7	.210	41	R40050	-----	280	5303UE
18/6	31UR.186	7	.226	52	R40015	E1036S	-----	5304UE
18/7	31UR.187	7	.240	56	R40051	-----	281	5305UE
18/8	31UR.188	7	.248	59	R40052	E1038S	-----	5306UE
18/9	31UR.189	7	.280	72	R40045	-----	282	5307UE
18/10	31UR.1810	7	.294	74	R40054	E1040S	-----	5308UE
18/12	31UR.1812	7	.316	93	R40024	E1041S	283	5309UE
18/15	31UR.1815	7	.351	115	R40058	-----	284	-----
18/20	31UR.1820	7	.400	152	R40063	-----	285	530BUE

Table 31UR-5C-16AWG Str. BC; PVC or PP Insul.; Cabled; Unshielded; Riser Rated

Const	SDN	A	B	C	TAPPAN	GENERAL CABLE	BELDEN-WEST PENN	BELDEN
16/5	31UR.165	19	.233	56	R50023	-----	-----	-----
16/6	31UR.166	19	.252	67	R50024	-----	-----	-----
16/7	31UR.167	19	.265	77	R50025	-----	-----	5205UE
16/8	31UR.168	19	.291	87	R50026	-----	-----	-----
16/10	31UR.1610	19	.313	107	R50027	-----	-----	-----
16/12	31UR.1612	19	.339	127	R50029	-----	-----	-----

Manufacturer's makes vary. Diameters (Col. B) and weights (Col. C) are conservative and for engineering purposes only. Consult manufacturer's data for individual make & model information.

Column A: Stranded bare copper conductor number of strands.
Column B: Nominal Cable Diameter in inches, for conduit fill calculations.
Column C: Nominal cable weight in pounds per 1000ft.

31UP: 300V Specification 31: Stranded Bare Copper; Cabled **Unshielded**
Str. BC / LS-PVC / Cabled / Unshielded (U) / Appropriate Jacket. **Plenum Rated (P).**

5/C and Higher Cables

Conductor: Stranded Bare Copper per Table 31UP-5C Column A

Insulation: Low Smoke PVC, thickness according to UL 444 or UL 13 as applicable. Fluoropolymer insulation may be used if needed for the particular construction to pass NFPA 262 flame test per manufacture's listing.

Color Code: Black, Red, White, Green, Brown, Blue, Orange, Yellow, Violet, Gray, Pink, Beige, White striped Black, White striped Red, White striped Green, White striped Orange, White striped Blue, White striped Brown, White striped Yellow, White striped Violet.

Assembly: Multiple conductors twisted together for flexibility. Ripcord under the jacket to facilitate removal.

Shielding: None.

Jacket: Low Smoke PVC. Thickness according to UL 444 or UL 13 as applicable. Fluoropolymer jacket material may be used if needed for the particular construction to pass NFPA 262 per manufacturer's listing. Color: white, natural; or gray unless specified.

Diameter: Nominal diameter shall not exceed that in table 31UP-5C Column B.

Flame test: Passes NFPA 262 (UL 910 Steiner Tunnel) flame test for use in plenum spaces.

Voltage: 300V but not printed on the jacket per UL 444 and UL 13. / 150V if listed CL2P only.

Application: 22–16 AWG: Communication Cable per NEC Art. 800 and UL 444 ;
All sizes: Class 3 or Class 2 Remote-Control, Signaling, and Power-Limited Circuits per NEC Art. 725 and UL 13.

Installation: Suitable for open cabling use in plenum spaces as CMP, CL3P or CL2P; also for open cabling use in risers and in horizontal runs since rated higher than those requirements.

Temp.: 0C to 60C minimum range, continuous operating ambient temperature.

QA: UL Listed as cable type CMP, CL3P, or both. Cable rating(s) shall be printed on the jacket. UL file number will be printed on the jacket alongside the gauge size and cable rating. UL label shall be displayed on every cable package near manufacturer's product label.

Packaging: 1000ft or 500ft: spools or reels. Boxes optional where practicable.

Appr. Manf: Manufacturer make and model shall be as listed in table 31UP-5C.

Availability: Cable shall be available from local distribution known to project management

Table 31UP-5C

Table 31UP-5C-22AWG Str. BC; LS-PVC Insul.; Cabled, Unshielded; Plenum Rated

Const	SDN	A	B	C	TAPPAN	GENERAL CABLE	BELDEN-WEST PENN	BELDEN
22/5	31UP.225	7	.162	22	P20135	-----	25855B	-----
22/6	31UP.226	7	.191	28	P20008	E3006S	25270B	6504UE
22/7	31UP.227	7	.163	25	P20137	-----	-----	-----
22/8	31UP.228	7	.206	34	P20018	E3008S	25271B	6506UE
22/9	31UP.229	7	.183	30	P20139	-----	-----	-----
22/10	31UP.2210	7	.218	43	P20140	E3010S	25272B	6508UE
22/12	31UP.2212	7	.225	45	P20044	E3012S	25273B	6509UE
22/15	31UP.2215	7	.246	53	P20145	-----	25274B	-----
22/20	31UP.2220	7	.259	62	P20150	-----	-----	-----

Table 31UP-5C-20AWG Str. BC; LS-PVC Insul.; Cabled, Unshielded; Plenum Rated

Const	SDN	A	B	C	TAPPAN	GENERAL CABLE	BELDEN-WEST PENN	BELDEN
20/5	31UP.205	7	.176	28	P30002	-----	-----	6403UE
20/6	31UP.206	7	.172	30	P30049	-----	-----	-----
20/7	31UP.207	7	.181	34	P30050	-----	-----	-----
20/8	31UP.208	7	.209	44	P30051	-----	-----	6406UE
20/9	31UP.209	7	.204	43	P30052	-----	-----	-----
20/10	31UP.2010	7	.214	47	P30053	-----	-----	-----
20/12	31UP.2012	7	.231	56	P30055	-----	-----	-----
20/15	31UP.2015	7	.255	69	P30056	-----	-----	-----
20/20	31UP.2020	7	.294	91	P30057	-----	-----	-----

Table 31UP-5C-18AWG Str. BC; LS-PVC Insul.; Cabled, Unshielded; Plenum Rated

Const	SDN	A	B	C	TAPPAN	GENERAL CABLE	BELDEN-WEST PENN	BELDEN
18/5	31UP.185	7	.203	40	P40163	C3125	-----	6303UE
18/6	31UP.186	7	.222	51	P40065	E3036S	25186B	6304UE
18/7	31UP.187	7	.217	50	P40164	-----	-----	-----
18/8	31UP.188	7	.242	65	P40030	E3038S	25188B	6306UE
18/9	31UP.189	7	.262	68	P40165	-----	-----	6307UE
18/10	31UP.1810	7	.286	75	P40166	C3123	-----	6308UE
18/12	31UP.1812	7	.296	84	P40168	C3124	-----	6309UE
18/15	31UP.1815	7	.308	103	P40171	-----	-----	-----
18/20	31UP.1820	7	.352	135	P40176	-----	-----	-----

Table 31UP-5C-16AWG Str. BC; LS-PVC Insul.; Cabled, Unshielded; Plenum Rated

Const	SDN	A	B	C	TAPPAN	GENERAL CABLE	BELDEN-WEST PENN	BELDEN
16/5	31UP.165	19	.221	56	P50014	-----	-----	-----
16/6	31UP.166	19	.240	66	P50015	-----	-----	-----
16/7	31UP.167	19	.253	76	P50016	-----	-----	-----
16/8	31UP.168	19	.279	87	P50017	-----	-----	-----
16/9	31UP.169	19	.287	97	P50018	-----	-----	-----
16/10	31UP.1610	19	.301	107	P50019	-----	-----	-----
16/12	31UP.1612	19	.327	128	P50021	-----	-----	-----

Manufacturer's makes vary. Diameters (Col. B) and weights (Col. C) are conservative and for engineering purposes only. Consult manufacturer's data for individual make & model information.

Column A: Stranded bare copper conductor number of strands.
Column B: Nominal Cable Diameter in inches, for conduit fill calculations.
Column C: Nominal cable weight in pounds per 1000ft.

31SR: 300V Specification 31: Stranded Bare Copper; Cabled **Shielded**
Str. BC / PVC or PP Insulation / Cabled / Shielded (S) / PVC Jacket. **Riser Rated (R).**

5/C and Higher Cables

Conductor: Stranded Bare Copper per Table 31SR-5C Column A

Insulation: PVC or Polypropylene, thickness according to UL 444 or UL 13 as applicable

Color Code: Black, Red, White, Green, Brown, Blue, Orange, Yellow, Violet, Gray, Pink, Beige, White striped Black, White striped Red, White striped Green, White striped Orange, White striped Blue, White striped Brown, White striped Yellow, White striped Violet.

Assembly: Multiple conductors twisted together for flexibility. Ripcord under the jacket to facilitate removal.

Shielding: Aluminum-polyester shielding tape, Tinned Copper Drain Wire. Aluminum foil surface facing out; drain wire is outside the tape and in continuous contact with the foil.

Jacket: PVC. Thickness according to UL 444 or UL 13 as applicable. Color: Gray unless specified.

Diameter: Nominal diameter shall not exceed that in table 31SR-5C Column B.

Flame test: Passes UL 1666, for use in risers.

Voltage: 300V but not printed on the jacket per UL 444 and UL 13. / 150V for items listed CL2R only.

Application: 22–16 AWG: Communication Cable per NEC Art. 800 and UL 444; All sizes: Class 3 or Class 2 Remote-Control, Signaling, and Power-Limited Circuits per NEC Art. 725 and UL 13.

Installation: Suitable for open cabling use in Risers as CMR, CL3R, or CL2R; also for open cabling horizontal runs since rated higher than CM, CL3, CL2 respectively. Not for use in plenum spaces unless running in conduit.

Temp.: 0C to 60C minimum range, continuous operating ambient temperature.

QA: UL Listed as cable type CMR (22-16AWG), CL3R, or both; rating(s) printed on the jacket. UL file number will be printed on the jacket alongside the gauge size and cable rating. UL label shall be displayed on every cable package near manufacturer's product label.

Packaging: 1000ft or 500ft: spools or reels. Boxes optional where practicable.

Appr. Manf: Manufacturer make and model shall be as listed in table 31SR-5C.

Availability: Cable shall be available from local distribution known to project management

Table 31SR-5C

Table 31SR-5C-22AWG Str. BC; PVC or PP Insul.; Cabled; Shielded, Riser Rated

Const	SDN	A	B	C	TAPPAN	GENERAL CABLE	BELDEN-WEST PENN	BELDEN
22/5	31SR.225	7	.172	23	R20775	-----	3855	5503FE
22/6	31SR.226	7	.187	29	R20076	E2006S	3270	5504FE
22/7	31SR.227	7	.196	29	R20077	-----	------	-----
22/8	31SR.228	7	.212	32	R20024	E2008S	3271	5506FE
22/9	31SR.229	7	.200	34	R20078	-----	------	-----
22/10	31SR.2210	7	.247	41	R20488	E2010S	3272	5508FE
22/12	31SR.2212	7	.255	47	R20081	E2012S	3273	-----
22/15	31SR.2215	7	.282	58	R20084	-----	3274	-----
22/20	31SR.2220	7	.312	73	R21152	-----	3275	-----

Table 31SR-5C-20AWG Str. BC; PVC or PP Insul.; Cabled; Shielded, Riser Rated

Const	SDN	A	B	C	TAPPAN	GENERAL CABLE	BELDEN-WEST PENN	BELDEN
20/5	31SR.205	7	.193	30	R30018	-----	3262	5403FE
20/6	31SR.206	7	.195	32	R30044	-----	-----	-----
20/7	31SR.207	7	.211	40	R30045	-----	3263	5405FE
20/8	31SR.208	7	.215	43	R30046	-----	-----	-----
20/9	31SR.209	7	.256	52	R30047	-----	3264	5407FE
20/10	31SR.2010	7	.258	54	R30048	-----	-----	-----
20/12	31SR.2012	7	.288	62	R30025	-----	3265	-----
20/15	31SR.2015	7	.290	72	R30050	-----	-----	-----
20/20	31SR.2020	7	.302	93	R30051	-----	-----	-----

Table 31SR-5C-18AWG Str. BC; PVC or PP Insul.; Cabled; Shielded, Riser Rated

Const	SDN	A	B	C	TAPPAN	GENERAL CABLE	BELDEN-WEST PENN	BELDEN
18/5	31SR.185	7	.214	45	R40070	-----	3280	5303FE
18/6	31SR.186	7	.243	57	R40611	E2036S	3021	5304FE
18/7	31SR.187	7	.245	60	R40071	-----	3281	5305FE
18/8	31SR.188	7	.270	64	R40014	E2038S	-----	5306FE
18/9	31SR.189	7	.284	72	R40073	-----	3282	5307FE
18/10	31SR.1810	7	.294	74	R40098	E2040S	-----	-----
18/12	31SR.1812	7	.319	93	R40075	E2041S	3283	-----
18/15	31SR.1815	7	.325	110	R40078	-----	3266	-----
18/20	31SR.1820	7	.395	150	R41151	-----	3285	-----

Manufacturer's makes vary. Diameters (Col. B) and weights (Col. C) are conservative and for engineering purposes only. Consult manufacturer's data for individual make & model information.

Column A: Stranded bare copper conductor number of strands.
Column B: Nominal Cable Diameter in inches, for conduit fill calculations.
Column C: Nominal cable weight in pounds per 1000ft.

31SP: 300V Specification 31: Stranded Bare Copper; Cabled Shielded
Str. BC / LS-PVC Insulation / Cabled / Shielded (S) / Appropriate Jacket. Plenum Rated (P).

5/C and Higher Cables

Conductor: Stranded Bare Copper per Table 31SP-5C Column A

Insulation: Low Smoke PVC, thickness according to UL 444 or UL 13 as applicable. Fluoropolymer insulation may be used if needed for the particular construction to pass NFPA 262.

Color Code: Black, Red, White, Green, Brown, Blue, Orange, Yellow, Violet, Gray, Pink, Beige, White striped Black, White striped Red, White striped Green, White striped Orange, White striped Blue, White striped Brown, White striped Yellow, White striped Violet.

Assembly: Multiple conductors twisted together for flexibility.
Ripcord under the jacket to facilitate removal.

Shielding: Aluminum-polyester shielding tape, Tinned Copper Drain Wire. Aluminum foil surface facing out; drain wire is outside the tape and in continuous contact with the foil.

Jacket: Low Smoke PVC. Thickness according to UL 444 or UL 13 as applicable. Fluoropolymer jacket material may be used if needed for the particular construction to pass NFPA 262 per mfg. listing. Color: White, natural, or gray unless specified.

Diameter: Nominal diameter shall not exceed that in table 31SP-5C Column B.

Flame test: Passes NFPA 262. (Formerly UL 910.)

Voltage: 300V but not printed on the jacket per UL 444 and UL 13. / 150V for items listed CL2P only.

Application: 22–16 AWG: Communication Cable per NEC Art. 800 and UL 444;
All sizes: Class 3 or Class 2 Remote-Control, Signaling, and Power-Limited Circuits per NEC Art. 725 and UL 13.

Installation: Suitable for open cabling use in plenum spaces as CMP, CL3P or CL2P; also for open cabling use in risers and in horizontal runs since rated higher than those requirements.

Temp.: 0C to 60C minimum range, continuous operating ambient temperature.

QA: UL Listed as cable type CMP, CL3P, or both. Cable rating(s) shall be printed on the jacket.
UL file number will be printed on the jacket alongside the gauge size and cable rating.
UL label shall be displayed on every cable package near manufacturer's product label.

Packaging: 1000ft or 500ft: spools or reels. Boxes optional where practicable.

Appr. Manf: Manufacturer make and model shall be as listed in table 31SP-5C.

Availability: Cable shall be available from local distribution known to project management

Table 31SP-5C

Table 31SP-5C-22AWG Str. BC; LS-PVC Insul.; Cabled; Shielded; Plenum Rated

Const	SDN	A	B	C	TAPPAN	GENERAL CABLE	BELDEN-WEST PENN	BELDEN
22/5	31SP.225	7	.165	22	P20899	-----	253855B	-----
22/6	31SP.226	7	.191	28	P20019	E2106S	253270B	6504FE
22/8	31SP.228	7	.206	35	P20037	E2108S	253271B	6506FE
22/9	31SP.229	7	.218	32	P20109	-----	-----	-----
22/10	31SP.2210	7	.229	43	P20022	-----	253272B	6508FE
22/12	31SP.2212	7	.246	45	P20023	-----	-----	-----
22/15	31SP.2215	7	.264	53	P20114	-----	253274B	-----
22/20	31SP.2220	7	.259	65	P20119	-----	-----	-----

Table 31SP-5C-20AWG Str. BC; LS-PVC Insul.; Cabled; Shielded; Plenum Rated

Const	SDN	A	B	C	TAPPAN	GENERAL CABLE	BELDEN-WEST PENN	BELDEN
20/5	31SP.205	7	.163	28	P30044	-----	-----	-----
20/6	31SP.206	7	.172	32	P30017	-----	-----	-----
20/7	31SP.207	7	.196	41	P30061	-----	-----	6405FE
20/8	31SP.208	7	.209	44	P30062	-----	-----	-----
20/9	31SP.209	7	.229	52	P30063	-----	-----	6407FE
20/10	31SP.2010	7	.214	47	P30064	-----	-----	-----

Table 31SP-5C-18AWG Str. BC; LS-PVC Insul.; Cabled; Shielded; Plenum Rated

Const	SDN	A	B	C	TAPPAN	GENERAL CABLE	BELDEN-WEST PENN	BELDEN
18/5	31SP.185	7	.195	40	P40135	-----	-----	-----
18/6	31SP.186	7	.232	52	P40040	E2206S	253186B	6304FE
18/8	31SP.188	7	.256	66	P40070	E2208S	253188B	6306FE
18/9	31SP.189	7	.266	73	P40137	-----	-----	6307FE
18/10	31SP.1810	7	.286	75	P40041	-----	-----	-----
18/12	31SP.1812	7	.306	99	P40139	-----	-----	6309FE

Table 31SP-5C-16AWG Str. BC; LS-PVC Insul.; Cabled; Shielded; Plenum Rated

Const	SDN	A	B	C	TAPPAN	GENERAL CABLE	BELDEN-WEST PENN	BELDEN
16/5	31SP.165	19	.221	58	P50024	-----	-----	-----
16/6	31SP.166	19	.240	70	P50025	-----	-----	-----
16/7	31SP.167	19	.253	80	P50026	-----	-----	-----
16/8	31SP.168	19	.279	90	P50031	-----	-----	-----

Manufacturer's makes vary. Diameters (Col. B) and weights (Col. C) are conservative and for engineering purposes only. Consult manufacturer's data for individual make & model information.

Column A: Stranded bare copper conductor number of strands.
Column B: Nominal Cable Diameter in inches, for conduit fill calculations.
Column C: Nominal cable weight in pounds per 1000ft.

32UR: 300V Specification 32: Stranded Bare Copper; Pairs (N/S) Unshielded O/A
Str. BC / PVC or PP Insul. / Multiple Pairs (N/S) / Unshielded (U) / PVC Jacket. Riser Rated (R).

Conductor:	Stranded Bare Copper. – 7 strands. Table 32UR Column A.
Insulation:	PVC or Polypropylene, thickness according to UL 444.
Color Code:	Pair (1) Black& Red, (2) Black&White, (3) Black&Green, (4) Black&Blue, (5) Black&Yellow, (6) Black&Brown, (7) Black&Orange, (8) Red&White, (9) Red&Green preferred. Other color coding methods as per Table 32UR manufacturers' makes and models if unavoidable.
Assembly:	Insulated conductors twisted into pairs. Multiple pairs cabled together. Ripcord under the jacket to facilitate removal.
Shielding:	None.
Jacket:	PVC. Thickness according to UL 444. Color: Gray unless specified.
Diameter:	Nominal diameter shall not exceed that in table 32UR Column B.
Flame test:	Passes UL 1666, for use in risers.
Voltage:	300V but not printed on the jacket per UL 444.
Application:	Communication Cable per NEC Art. 800 and UL 444; also for Class 3 or Class 2 Remote-Control, Signaling, and Power-Limited Circuits per NEC Art. 725.
Installation:	Suitable for open cabling use in Risers as CMR, CL3R, or CL2R; also for open cabling horizontal runs since rated higher than CM, CL3, CL2 respectively. Not for use in plenum spaces unless running in conduit.
Temp.:	0C to 60C minimum range, continuous operating ambient temperature.
QA:	UL Listed as cable type CMR. Cable ratings shall be printed on the jacket. UL file number will be printed on the jacket alongside the gauge size and cable rating. UL label shall be displayed on every cable package near manufacturer's label.
Packaging:	1000ft or 500ft: spools, reels or boxes.
Appr. Manf:	Manufacturer make and model shall be as listed in table 32UR.
Availability:	Cable shall be available from local distribution known to project management.

Table 32UR appears on page 13410-46

32UP: 300V Specification 32: Stranded Bare Copper; Pairs (N/S) Unshielded O/A
Str. BC / LS-PVC Insul. / Multiple Pairs (N/S) / Unshielded (U) / Suitable Jacket. Plenum Rated (P).

Conductor: Stranded Bare Copper. – 7 strands. Table 32UP Column A.

Insulation: LS-PVC preferred. Fluoropolymer if required to pass NFPA 262 flame and smoke test. Manufacturer's makes may vary. Thickness according to UL 444.

Color Code: Pair (1) Black& Red, (2) Black&White, (3) Black&Green, (4) Black&Blue, (5) Black&Yellow, (6) Black&Brown, (7) Black&Orange, (8) Red&White, (9) Red&Green preferred. Other color coding methods as per Table 32UP manufacturers' makes and models if unavoidable..

Assembly: Insulated conductors twisted into pairs. Multiple pairs cabled together.
Ripcord under the jacket to facilitate removal.

Shielding: None.

Jacket: LS-PVC preferred. Fluoropolymer material if needed to pass NFPA 262 plenum flame and smoke test per manufacturer's listings.
Thickness according to UL 444. Color: Natural, white, or gray unless specified.

Diameter: Nominal diameter shall not exceed that in table 32UP Column B.

Flame test: Passes NFPA 262 (formerly UL 910,) for use in plenum spaces

Voltage: 300V but not printed on the jacket per UL 444.

Application: Communication Cable per NEC Art. 800 and UL 444; also for Class 3 or Class 2 Remote-Control, Signaling, and Power-Limited Circuits per NEC Art. 725.

Installation: Suitable for open cabling use in plenum spaces as CMP, CL3P, or CL2P; also for open cabling in risers and in horizontal runs since rated higher than those requirements.

Temp.: 0C to 60C minimum range, continuous operating ambient temperature.

QA: UL Listed as cable type CMP. Cable ratings shall be printed on the jacket.
UL file number will be printed on the jacket alongside the gauge size and cable rating.
UL label shall be displayed on every cable package near manufacturer's label.

Packaging: 1000ft or 500ft: spools, reels or boxes.

Appr. Manf: Manufacturer make and model shall be as listed in table 32UP.

Availability: Cable shall be available from local distribution known to project management.

Table 32UP appears on page 13410-46

32SR: 300V Specification 32: Stranded Bare Copper; Pairs (N/S) Shielded O/A
Str. BC / PVC or PP Insul. / Multiple Pairs (N/S) / Shielded Overall (S) / PVC Jacket. Riser Rated (R).

Conductor: Stranded Bare Copper. – 7 strands Table 32SR Column A.

Insulation: PVC or Polypropylene, thickness according to UL 444.

Color Code: Pair (1) Black& Red, (2) Black&White, (3) Black&Green, (4) Black&Blue, (5) Black&Yellow, (6) Black&Brown, (7) Black&Orange, (8) Red&White, (9) Red&Green preferred. Other color coding methods as per Table 32SR manufacturers' makes and models if unavoidable.

Assembly: Insulated conductors twisted into pairs. Multiple pairs cabled together.
Ripcord under the jacket to facilitate removal.

Shielding: Overall – Aluminum-polyester shielding tape, Tinned Copper Drain Wire. Aluminum foil surface facing out; drain wire is outside the tape and in continuous contact with the foil.

Jacket: PVC. Thickness according to UL 444. Color: gray unless specified.

Diameter: Nominal diameter shall not exceed that in table 32SR Column B.

Flame test: Passes UL 1666, for use in risers.

Voltage: 300V but not printed on the jacket per UL 444.

Application: Communication Cable per NEC Art. 800 and UL 444; also for Class 3 or Class 2 Remote-Control, Signaling, and Power-Limited Circuits per NEC Art. 725.

Installation: Suitable for open cabling use in Risers as CMR, CL3R, or CL2R; also for open cabling horizontal runs since rated higher than CM, CL3, CL2 respectively.
Not for use in plenum spaces unless running in conduit.

Temp.: 0C to 60C minimum range, continuous operating ambient temperature.

QA: UL Listed as cable type CMR. Cable ratings shall be printed on the jacket.
UL file number will be printed on the jacket alongside the gauge size and cable rating.
UL label shall be displayed on every cable package near manufacturer's label.

Packaging: 1000ft or 500ft: spools, reels or boxes.

Appr. Manf: Manufacturer make and model shall be as listed in table 32SR.

Availability: Cable shall be available from local distribution known to project management.

Table 32SR appears on page 13410-47

32SP: 300V Specification 32: Stranded Bare Copper; Pairs (N/S) Shielded O/A
Str. BC / LS-PVC Insul. / Multiple Pairs (N/S) / Shielded Overall (S) / Suitable Jacket. Plenum Rated (P).

Conductor: Stranded Bare Copper. – 7 strands. Table 32SP Column A.

Insulation: LS-PVC preferred. Fluoropolymer if required to pass NFPA 262 flame and smoke test per manufacturer's listing. Thickness according to UL 444.

Color Code: Pair (1) Black& Red, (2) Black&White, (3) Black&Green, (4) Black&Blue, (5) Black&Yellow, (6) Black&Brown, (7) Black&Orange, (8) Red&White, (9) Red&Green preferred. Other color coding methods as per Table 32SP manufacturers' makes and models if unavoidable.

Assembly: Insulated conductors twisted into pairs. Multiple pairs cabled together.
Ripcord under the jacket to facilitate removal.

Shielding: Overall – Aluminum-polyester shielding tape, Tinned Copper Drain Wire. Aluminum foil surface facing out; drain wire is outside the tape and in continuous contact with the foil.

Jacket: LS-PVC preferred. Fluoropolymer if required to pass NFPA 262 flame test per manufacturer's listing. Thickness according to UL 444. Color: Natural, white, or gray unless specified.

Diameter: Nominal diameter shall not exceed that in table 32SP Column B.

Flame test: Passes NFPA 262 (formerly UL 910) test for use in plenum spaces.

Voltage: 300V but not printed on the jacket per UL 444.

Application: Communication Cable per NEC Art. 800 and UL 444; also for Class 3 or Class 2 Remote-Control, Signaling, and Power-Limited Circuits per NEC Art. 725.

Installation: Suitable for open cabling use in plenum spaces CMP, CL3P, or CL2P; also for open cabling in risers and in horizontal runs since rated higher than those requirements.

Temp.: 0C to 60C minimum range, continuous operating ambient temperature.

QA: UL Listed as cable type CMP. Cable ratings shall be printed on the jacket.
UL file number will be printed on the jacket alongside the gauge size and cable rating.
UL label shall be displayed on every cable package near manufacturer's label.

Packaging: 1000ft or 500ft: spools, reels or boxes.

Appr. Manf: Manufacturer make and model shall be as listed in table 32SP.

Availability: Cable shall be available from local distribution known to project management.

Table 32SP appears on page 13410-47

Table 32UR Str. BC; PVC or PP Insul.; Multiple Pairs (N/S); Unshld O/A; Riser Rated

Const	SDN	A	B	C	TAPPAN	GENERAL CABLE	BELDEN-WEST PENN	BELDEN
22/2pr	32UR.222	7	.199	19	R29984	-----	651	5541UE
22/3pr	32UR.223	7	.212	26	R29983	-----	652	5542UE
22/4pr	32UR.224	7	.234	32	R29982	-----	653	5543UE
22/6pr	32UR.226	7	.289	50	R29980	-----	654	-----
22/9pr	32UR.229	7	.340	70	R29977	-----	655	5547UE
22/12pr	32UR.2212	7	.384	85	R29974	-----	656	-----
22/15pr	32UR.2215	7	.410	102	R29971	-----	-----	-----
22/16pr	32UR.2216	7	.428	110	R29970	-----	657	-----
18/2pr	32UR.182	7	.266	38	R49912	-----	751	5341UE
18/3pr	32UR.183	7	.283	50	R49992	-----	752	5342UE
18/4pr	32UR.184	7	.320	67	R49671	-----	753	5343UE
18/6pr	32UR.186	7	.366	96	R49914	-----	754	5345UE
18/9pr	32UR.189	7	.434	140	R49611	-----	755	5347UE
18/12pr	32UR.1812	7	.475	170	R49976	-----	------	-----

Manufacturer's makes vary. Diameters (Col. B) and weights (Col. C) are conservative and for engineering purposes only. Consult manufacturer's data for individual make & model information.

Column A: Stranded bare copper conductor number of strands.
Column B: Nominal Cable Diameter in inches, for conduit fill calculations.
Column C: Nominal cable weight in pounds per 1000ft.

Table 32UP Str. BC; LS-PVC Insul.; Multiple Pairs (N/S); Unshld O/A; Plenum Rated

Const	SDN	A	B	C	TAPPAN	GENERAL CABLE	BELDEN-WEST PENN	BELDEN
22/2pr	32UP.222	7	.200	20	P29966	-----	-----	6541UE
22/3pr	32UP.223	7	.212	27	P29965	-----	-----	6542UE
22/4pr	32UP.224	7	.234	34	P29803	-----	-----	6543UE
22/6pr	32UP.226	7	.264	47	P29962	-----	-----	6545UE
22/9pr	32UP.229	7	.318	69	P29959	-----	-----	6547UE
22/12pr	32UP.2212	7	.338	80	P29956	-----	-----	-----
22/15pr	32UP.2215	7	.365	97	P29953	-----	-----	-----
22/16pr	32UP.2216	7	.383	105	P29952	-----	-----	-----
18/2pr	32UP.182	7	.260	38	P49996	-----	25751B	6341UE
18/3pr	32UP.183	7	.278	50	P49861	-----	25752B	6342UE
18/4pr	32UP.184	7	.314	66	P49860	-----	25753B	6343UE
18/6pr	32UP.186	7	.355	95	P49864	-----	-----	6345UE
18/9pr	32UP.189	7	.423	138	-----	-----	-----	6347UE

Manufacturer's makes vary. Diameters (Col. B) and weights (Col. C) are conservative and for engineering purposes only. Consult manufacturer's data for individual make & model information.

Column A: Stranded bare copper conductor number of strands.
Column B: Nominal Cable Diameter in inches, for conduit fill calculations.
Column C: Nominal cable weight in pounds per 1000ft.

Table 32SR **Str. BC; PVC or PP Insul.; Multiple Pairs(N/S); Shield O/A; Riser Rated**

Const	SDN	A	B	C	TAPPAN	GENERAL CABLE	BELDEN-WEST PENN	BELDEN
22/2pr	32SR.222	7	.209	21	R29454	-----	3651	5541FE
22/3pr	32SR.223	7	.233	29	R29946	-----	3652	5542FE
22/4pr	32SR.224	7	.246	40	R29569	-----	3653	5543FE
22/6pr	32SR.226	7	.288	55	R29943	-----	3654	5545FE
22/9pr	32SR.229	7	.300	68	R29989	-----	------	------
22/12pr	32SR.2212	7	.356	85	R29563	-----	------	------
22/15pr	32SR.2215	7	.383	102	R29934	-----	------	------
22/16pr	32SR.2216	7	.401	109	R29933	-----	------	------
18/2pr	32SR.182	7	.270	40	R49668	-----	3751	5341FE
18/3pr	32SR.183	7	.281	51	R49988	-----	3752	5342FE
18/4pr	32SR.184	7	.318	70	R49670	-----	3753	5343FE
18/6pr	32SR.186	7	.373	101	R49985	-----	------	5345FE

Manufacturer's makes vary. Diameters (Col. B) and weights (Col. C) are conservative and for engineering purposes only. Consult manufacturer's data for individual make & model information.

Column A: Stranded bare copper conductor number of strands.
Column B: Nominal Cable Diameter in inches, for conduit fill calculations.
Column C: Nominal cable weight in pounds per 1000ft.

Table 32SP **Str. BC; LS-PVC Insul.; Multiple Pairs (N/S); ShieldO/A; Plenum Rated**

Const	SDN	A	B	C	TAPPAN	GENERAL CABLE	BELDEN-WEST PENN	BELDEN
22/2pr	32SP.222	7	.203	22	P29795	------	------	6541FE
22/3pr	32SP.223	7	.216	28	P29689	------	------	6542FE
22/4pr	32SP.224	7	.238	35	P29704	------	------	6543FE
22/6pr	32SP.226	7	.268	48	P29929	------	------	6545FE
22/9pr	32SP.229	7	.323	71	P29926	------	------	6547FE
22/12pr	32SP.2212	7	.338	83	P29924	------	------	------
22/15pr	32SP.2215	7	.365	101	P29923	------	------	------
18/2pr	32SP.182	7	.183	34	P49873	------	------	------
18/3pr	32SP.183	7	.282	50	P49867	------	------	6342FE
18/4pr	32SP.184	7	.318	73	P49859	------	------	6343FE
18/6pr	32SP.186	7	.359	100	P49863	------	------	6345FE
18/9pr	32SP.189	7	.427	143	-----	------	------	6347FE

Manufacturer's makes vary. Diameters (Col. B) and weights (Col. C) are conservative and for engineering purposes only. Consult manufacturer's data for individual make & model information.

Column A: Stranded bare copper conductor number of strands.
Column B: Nominal Cable Diameter in inches, for conduit fill calculations.
Column C: Nominal cable weight in pounds per 1000ft.

33UR: 300V Specification 33: Stranded Bare Copper; Pairs, EPS/CD Unshielded O/A
Str. BC / PVC or PP Insul. / Multiple Pairs (EPS/CD) / Unshielded O/A (U) / PVC Jkt. Riser Rated (R).

Each Pair Shielded (EPS) – Common Drain Wire (CD)

Conductor: Stranded Bare Copper. – 7 strands. Table 33UR Column A.

Insulation: PVC or Polypropylene, thickness according to UL 444.

Color Code: Pair (1) Black& Red, (2) White & Green

Assembly: Insulated conductors twisted into pairs and shielded (foil side out.). Multiple pairs cabled together with a common drain wire. Ripcord for jacket removal. PVC Jacket.

Shielding: Each pair: Aluminum-polyester shielding tape. Aluminum foil surface facing out. Tinned Copper Drain Wire. Drain wire is outside the tapes and in continuous contact with the both foil shields.

Jacket: PVC. Thickness according to UL 444. Color: Gray unless specified.

Diameter: Nominal diameter shall not exceed that in table 33UR Column B.

Flame test: Passes UL 1666, for use in risers.

Voltage: 300V but not printed on the jacket per UL 444..

Application: Communication Cable per NEC Art. 800 and UL 444; also for Class 3 or Class 2 Remote-Control, Signaling, and Power-Limited Circuits per NEC Art. 725.

Installation: Suitable for open cabling use in Risers as CMR, CL3R, or CL2R; also for open cabling horizontal runs since rated higher than CM, CL3, CL2 respectively.
Not for use in plenum spaces unless running in conduit.

Temp.: 0C to 60C minimum range, continuous operating ambient temperature.

QA: UL Listed as cable type CMR. Cable ratings shall be printed on the jacket.
UL file number will be printed on the jacket alongside the gauge size and cable rating.
UL label shall be displayed on every cable package near manufacturer's label.

Packaging: 1000ft or 500ft: spools, reels or boxes.

Appr. Manf: Manufacturer make and model shall be as listed in table 33UR.

Availability: Cable shall be available from local distribution known to project management.

Table 33UR Str. BC; PVC or PP Insul.; Pairs, EPS/CD; Overall N/S; Riser Rated

Const	SDN	A	B	C	TAPPAN	Alternatives: Specifications 323; Stranded Tinned Copper, Upgrade Insul. Table 323UH
22/2pr	33UR.222	7	.160	18	R29575	
18/2pr	33UR.182	7	.198	34	R49620	

Manufacturer's makes vary. Diameters (Col. B) and weights (Col. C) are conservative and for engineering purposes only. Consult manufacturer's data for individual make & model information.

Column A: Stranded bare copper conductor number of strands.
Column B: Nominal Cable Diameter in inches, for conduit fill calculations.
Column C: Nominal cable weight in pounds per 1000ft.

33UP: 300V Specification 33: Stranded Bare Copper; Pairs, EPS/CD Unshielded O/A
Str. BC / LS-PVC Insul. / Multiple Pairs (EPS/CD) / Unshielded O/A (U) / Jacket. Plenum Rated (P).

Each Pair Shielded (EPS) – Common Drain Wire (CD)

Conductor: Stranded Bare Copper. – 7 strands. Table 33UP Column A.

Insulation: LS-PVC, thickness according to UL 444.

Color Code: Pair (1) Black& Red, (2) White & Green.

Assembly: Insulated conductors twisted into pairs and shielded (foil side out.). Multiple pairs cabled together with a common drain wire. Ripcord for jacket removal. PVC Jacket.

Shielding: Each pair: Aluminum-polyester shielding tape. Aluminum foil surface facing out. Tinned Copper Drain Wire. Drain wire is outside the tapes and in continuous contact with the both foil shields.

Jacket: LS-PVC. Thickness according to UL 444. Color: White unless specified.

Diameter: Nominal diameter shall not exceed that in table 33UP Column B.

Flame test: NFPA 262, plenum cable test.

Voltage: 300V but not printed on the jacket per UL 444..

Application: Communication Cable per NEC Art. 800 and UL 444; also for Class 3 or Class 2 Remote-Control, Signaling, and Power-Limited Circuits per NEC Art. 725.

Installation: Suitable for open cabling use in plenum spaces. Also for open cabling in risers or horizontal runs since rated higher than those requirements.

Temp.: 0C to 60C minimum range, continuous operating ambient temperature.

QA: UL Listed as cable type CMP. Cable ratings shall be printed on the jacket.
UL file number will be printed on the jacket alongside the gauge size and cable rating.
UL label shall be displayed on every cable package near manufacturer's label.

Packaging: 1000ft or 500ft: spools, reels or boxes.

Appr. Manf: Manufacturer make and model shall be as listed in table 33UP.

Availability: Cable shall be available from local distribution known to project management.

Table 33UP Str. BC; LS-PVC Insul.; Pairs, EPS/CD; Overall N/S; Plenum Rated

Const	SDN	A	B	C	TAPPAN	GENERAL	BELDEN-WEST PENN	BELDEN
22/2pr	33UP.222	7	.204	18	P29841	-----	25510B	-----
18/2pr	33UP.182	7	.257	35	P49995	-----	25440B	-----

Manufacturer's makes vary. Diameters (Col. B) and weights (Col. C) are conservative and for engineering purposes only. Consult manufacturer's data for individual make & model information.

Column A: Stranded bare copper conductor number of strands.
Column B: Nominal Cable Diameter in inches, for conduit fill calculations.
Column C: Nominal cable weight in pounds per 1000ft.

34UR: 300V Specification 34: Stranded Bare Copper; Pairs, EPS Unshielded O/A
Str. BC / PVC or PP Insul. / Multiple Pairs (EPS) / Unshielded O/A (U) / PVC Jacket. Riser Rated (R).

Each Pair Shielded (EPS) – Isolated shields, one drain wire per pair.

Conductor: Stranded Bare Copper. – 7 strands. Table 34UR Column A.

Insulation: PVC or Polypropylene, thickness according to UL 444.

Color Code: Pair (1) Black& Red, (2) Black&White, (3) Black&Green, (4) Black&Blue, (5) Black&Yellow, (6) Black&Brown, (7) Black&Orange, (8) Red&White, (9) Red&Green. Etc. The black or red (first) wire striped with the color of the other leg in that pair. For color code above 9 pairs consult manufactures' specification.

Assembly: Insulated conductors twisted into pairs with a drain wire and tape shield (foil side in.) Multiple pairs cabled together. Shields are electrically isolated from each other.
Ripcord for jacket removal. PVC Jacket.

Shielding: Each pair: Aluminum-polyester shielding tape and a tinned copper drain wire. Aluminum foil surface facing in. Tape will be made with foil-free edges to ensure shield isolation or separate isolation tapes will be used. Overall: no overall shield.

Jacket: PVC. Thickness according to UL 444. Color: Gray unless specified.

Diameter: Nominal diameter shall not exceed that in table 34UR Column B.

Flame test: Passes UL 1666, for use in risers.

Voltage: 300V but not printed on the jacket per UL 444.

Application: Communication Cable per NEC Art. 800 and UL 444; also for Class 3 or Class 2 Remote-Control, Signaling, and Power-Limited Circuits per NEC Art. 725.

Installation: Suitable for open cabling use in Risers as CMR, CL3R, or CL2R; also for open cabling horizontal runs since rated higher than CM, CL3, CL2 respectively.
Not for use in plenum spaces unless running in conduit.

Temp.: 0C to 60C minimum range, continuous operating ambient temperature.

QA: UL Listed as cable type CMR. Cable ratings shall be printed on the jacket.
UL file number will be printed on the jacket alongside the gauge size and cable rating.
UL label shall be displayed on every cable package near manufacturer's label.

Packaging: 1000ft or 500ft: spools, reels or boxes.

Appr. Manf: Manufacturer make and model shall be as listed in table 34UR.

Availability: Cable shall be available from local distribution known to project management.

Table 34UR appears on 13410-52

34UP: 300V Specification 34: Stranded Bare Copper; Pairs, EPS **Unshielded O/A**
Str. BC / LS-PVC Insul. / Multiple Pairs (EPS) / Unshielded O/A (U) / Jacket. **Plenum Rated (P).**

Each Pair Shielded (EPS) – Isolated shields, one drain wire per pair.

Conductor: Stranded Bare Copper. – 7 strands. Table 34UP Column A.

Insulation: LS-PVC, thickness according to UL 444.

Color Code: Pair (1) Black& Red, (2) Black&White, (3) Black&Green, (4) Black&Blue, (5) Black&Yellow, (6) Black&Brown, (7) Black&Orange, (8) Red&White, (9) Red&Green. Etc.
For color code above 9 pairs consult manufacturer's specification.

Assembly: Insulated conductors twisted into pairs with a drain wire and tape shield (foil side in.) Multiple pairs cabled together. Shields are electrically isolated from each other.
Ripcord for jacket removal. PVC Jacket.

Shielding: Each pair: Aluminum-polyester shielding tape and a tinned copper drain wire. Aluminum foil surface facing in. Tape will be made with foil-free edges to ensure shield isolation or separate isolation tapes will be used. Overall: no overall shield.

Jacket: LS-PVC preferred. Fluoropolymer if required to pass NFPA 262 flame test per manufacturer's listing. Thickness according to UL 444.
Color: Natural, white, or gray unless specified.

Diameter: Nominal diameter shall not exceed that in table 34UP Column B.

Flame test: Passes NFPA 262 for use in plenum spaces.

Voltage: 300V but not printed on the jacket per UL 444.

Application: Communication Cable per NEC Art. 800 and UL 444; also for Class 3 or Class 2 Remote-Control, Signaling, and Power-Limited Circuits per NEC Art. 725.

Installation: Suitable for open cabling use in plenum spaces CMP, CL3P, or CL2P; also for open cabling in risers and in horizontal runs since rated higher than those requirements.

Temp.: 0C to 60C minimum range, continuous operating ambient temperature.

QA: UL Listed as cable type CMP. Cable ratings shall be printed on the jacket.
UL file number will be printed on the jacket alongside the gauge size and cable rating.
UL label shall be displayed on every cable package near manufacturer's label.

Packaging: 1000ft or 500ft: spools, reels or boxes.

Appr. Manf: Manufacturer make and model shall be as listed in table 34UP.

Availability: Cable shall be available from local distribution known to project management.

Table 34UP appears on 13410-52

Table 34UR **Str. BC; PVC or PP Insul.; EPS; Overall N/S; Riser Rated**

Const	SDN	A	B	C	TAPPAN
22/2pr	34UR.222	7	.160	21	R29840
22/3pr	34UR.223	7	.190	30	R29564
22/4pr	34UR.224	7	.252	40	R29617
22/6pr	34UR.226	7	.270	56	R29520
22/9pr	34UR.229	7	.325	84	R29615
22/12pr	34UR.2212	7	.361	109	R29997
22/20pr	34UR.2220	7	.434	177	R29828
22/25pr	34UR.2225	7	.497	219	R29827
22/27pr	34UR.2227	7	.506	235	R29826
22/30pr	34UR.2230	7	.534	264	R29580
22/35pr	34UR.2235	7	.565	304	R29579
18/2pr	34UR.182	7	.198	37	R49910
18/3pr	34UR.183	7	.240	54	R49909
18/4pr	34UR.184	7	.356	81	R49597

Alternatives:
Specification 314;
Solid Bare Copper
Table 314UR

Or

Specification 324
Stranded Tinned Copper
Upgrade Insulation
Table 324UH

Manufacturer's makes vary. Diameters (Col. B) and weights (Col. C) are conservative and for engineering purposes only. Consult manufacturer's data for individual make & model information.

Column A: Stranded bare copper conductor number of strands.
Column B: Nominal Cable Diameter in inches, for conduit fill calculations.
Column C: Nominal cable weight in pounds per 1000ft.

Table 34UP **Str. BC; LS-PVC Insul.; EPS; Overall N/S; Plenum Rated**

Const	SDN	A	B	C	TAPPAN	BELDEN-WEST PENN
22/2pr	34UP.222	7	.148	22	P29691	-----
22/3pr	34UP.223	7	.178	30	P29996	-----
22/4pr	34UP.224	7	.240	41	P29716	-----
22/6pr	34UP.226	7	.278	58	P29671	-----
22/9pr	34UP.229	7	.307	86	P29824	-----
22/12pr	34UP.2212	7	.343	112	P29821	-----
22/15pr	34UP.2215	7	.370	143	P19979	-----
18/2pr	34UP.182	7	.186	38	P49852	-----
18/3pr	34UP.183	7	.228	55	P49994	25441B
18/4pr	34UP.184	7	.324	74	P49853	-----
18/6pr	34UP.186	7	.349	107	P49840	-----
18/9pr	34UP.189	7	.424	163	P49910	-----

Alternatives:
Specification 314;
Solid Bare Copper
Table 314UP

Or

Specification 324
Stranded Tinned Copper
Upgrade Insulation
Table 324UP

Manufacturer's makes vary. Diameters (Col. B) and weights (Col. C) are conservative and for engineering purposes only. Consult manufacturer's data for individual make & model information.

Column A: Stranded bare copper conductor number of strands.
Column B: Nominal Cable Diameter in inches, for conduit fill calculations.
Column C: Nominal cable weight in pounds per 1000ft.

35MR: 300V Specification 35: Stranded Bare Copper; 1 pr + #/C **1 Pr Shielded (+) #/C**
Str. BC / PVC or PP Insul. / 1 Pr Shielded + Unshld. Cond.'s (M) / PVC Jacket **Riser Rated (R).**
Mixed Shielded Cables – Combination 1 Shielded Pair + #/C.

Conductor: Stranded Bare Copper. – 7 strands. Unless indicated; per table 35MR Column A.

Insulation: PVC or Polypropylene, thickness according to UL 444.

Color Code: Black, Red (Inside the shield); White, Green outside the shield.

Assembly: Black, Red insulated conductors twisted into a pair with a drain wire and tape shield (foil side in.) White and green (if present) insulated wires outside the shield. Ripcord for jacket removal. PVC Jacket.

Shielding: On the pair: Aluminum-polyester shielding tape and a tinned copper drain wire. Aluminum foil surface facing in. No overall shield

Jacket: PVC. Thickness according to UL 444. Color: Gray unless specified.

Diameter: Nominal diameter shall not exceed that in table 35M Column B.

Flame test: Passes UL 1666, for use in risers.

Voltage: 300V but not printed on the jacket per UL 444..

Application: Communication Cable per NEC Art. 800 and UL 444; also for Class 3 or Class 2 Remote-Control, Signaling, and Power-Limited Circuits per NEC Art. 725.

Installation: Suitable for open cabling use in Risers as CMR, CL3R, or CL2R; also for open cabling horizontal runs since rated higher than CM, CL3, CL2 respectively.
Not for use in plenum spaces unless running in conduit.

Temp.: 0C to 60C minimum range, continuous operating ambient temperature.

QA: UL Listed as cable type CMR. Cable ratings shall be printed on the jacket.
UL file number will be printed on the jacket alongside the gauge size and cable rating.
UL label shall be displayed on every cable package near manufacturer's label.

Packaging: 1000ft or 500ft: spools, reels or boxes.

Appr. Manf: Manufacturer make and model shall be as listed in table 35MR.

Availability: Cable shall be available from local distribution known to project management.

Table 35MR **Str. BC*; PVC or PP Insul.; Mixed Shielded + N/S; Riser Rated**

Const	SDN	A	B	C	TAPPAN	BELDEN-WEST PENN	BELDEN
1pr 22 SH + 1C 22	35MR.223	7	.171	16	R20725	352	5501GE
1pr 22 SH + 2C 22	35MR.224	7	.186	19	R20008	357	5502GE
1pr 20 SH + 1C 20	35MR.203	7	.196	23	R30087	358	5401GE
1pr 20 SH + 2C 20	35MR.204	7	.200	26	R30001	359	5402GE
1pr 20 SH + 2C 18	35MR.M14	7	.200	35	H90042	356	-----
1 pr 18 SH +1C 18	35MR.183	7	.177	28	R40871	-----	-----
1pr 18 SH + 2C 18	35MR.184	7	.225	38	R40872	360	5302GE
1pr 18 (N/S) + 1pr 22 (N/S)	35MR.M24	7	.309	30	H91299*	374	-----
1pr 18 (N/S) + 1pr 22 SH	35MR.M34	7	.299	33	H90772	-----	-----

*Tinned Copper

Column A: Stranded bare (except where noted) copper conductor number of strands.
Column B: Nominal Cable Diameter in inches, for engineering purposes only.
Column C: Nominal cable weight in pounds per 1000ft, for engineering purposes only.

35MP: 300V Specification 35: Stranded Bare Copper; 1 pr + #/C **1 Pr Shielded (+) #/C**
Str. BC / LS-PVC Insul. / 1 Pr Shielded + Unshld. Conductors (M) / Jacket. **Plenum Rated (P).**

Mixed Shielded Cables – Combination 1 Shielded Pair + #/C.

Conductor: Stranded Bare Copper. – 7 strands. Table 35MP Column A.

Insulation: LS-PVC, thickness according to UL 444.

Color Code: Each wire within shall be a unique color. No colors may repeat. Refer to mfg.'s specification.

Assembly: Per table 35MP CONST column. Ripcord for jacket removal. LS-PVC Jacket.

Shielding: On the pair where indicated: Aluminum-polyester shielding tape and a tinned copper drain wire. Aluminum foil surface facing in. No overall shield unless indicated.

Jacket: LS-PVC. Thickness according to UL 444. Color: White or natural unless specified.

Diameter: Nominal diameter shall not exceed that in table 3MP Column B.

Flame test: Passes NFPA 262 for use in plenum spaces.

Voltage: 300V but not printed on the jacket per UL 444.

Application: Communication Cable per NEC Art. 800 and UL 444; also for Class 3 or Class 2 Remote-Control, Signaling, and Power-Limited Circuits per NEC Art. 725.

Installation: Suitable for open cabling use in plenums as CMP, CL3P, or CL2P; also for open cabling In risers and in horizontal runs since rated higher than those requirements.

Temp.: 0C to 60C minimum range, continuous operating ambient temperature.

QA: UL Listed as cable type CMP. Cable ratings shall be printed on the jacket. UL file number will be printed on the jacket alongside the gauge size and cable rating. UL label shall be displayed on every cable package near manufacturer's label.

Packaging: 1000ft or 500ft: spools, reels or boxes.

Appr. Manf: Manufacturer make and model shall be as listed in table 32UR.

Availability: Cable shall be available from local distribution known to project management.

Table 35MP Str. BC; LS-PVC Insul.; Mixed Shielded + N/S; Plenum Rated

Const	SDN	A	B	C	TAPPAN	BELDEN-WEST PENN	BELDEN
1pr 22 SH + 1C 22	35MP.223	7	.171	16	P20024	25352B	6501GE
1pr 22 SH + 2C 22	35MP.224	7	.186	19	P20012	25357B	6502GE
1pr 20 SH + 1C 20	35MP.203	7	.196	23	P30195	25358B	6401GE
1pr 20 SH + 2C 20	35MP.204	7	.200	26	P30011	25359B	-----
1pr 20 SH + 2C 18	35MP.M14	7	.200	35	H90270	-----	-----
1pr 18 SH + 2C 18	35MP.184	7	.225	38	P40028	-----	-----
1pr 18 N/S + 1pr 22 N/S	35MP.M24	7	.275	27	H91322	25374	-----
1pr 18 N/S + 4C 22 N/S	35MP.M46	7	.310	33	H91741	-----	-----
1pr 18 N/S + 4C 22NS - OAS	35MP.M56	7	.310	35	H91576	-----	-----
1pr 18 N/S + 1pr 20 N/S	35MP.M64	7	.354	41	H92035	-----	-----

Column A: Stranded bare copper conductor number of strands.
Column B: Nominal Cable Diameter in inches, for engineering purposes only.
Column C: Nominal cable weight in pounds per 1000ft for engineering purposes only.

PERIODIC TABLE OF PARTIAL CONTENTS – Detail for Stranded Bare Copper, UL minimum Insulation

Cabled; Stranded Bare Copper; UL Minimum Insulation **Specification 31**

Multiple Unshielded Pairs; Stranded Bare Copper; UL Minimum Insulation **Specification 32**

Multiple Pairs; EPS/CD; Stranded Bare Copper; UL Minimum Insulation **Specification 33**

Multiple Pairs, EPS; Stranded Bare Copper; UL Minimum Insulation **Specification 34**

Combo Cables; Str BC; UL Minimum Insulation; Mixed Shielding /AWG Sizes **Specification 35**

Access Control Combo Cables; Str BC; Mixed Constr. Individual Inner Jackets **Specification 36**

Audio and Speaker Cable Stranded Bare Copper, Cabled. In-Wall Speaker **Specification 27**

36MR: 300V Specification 36: Access Control Door Composite Cable **Card Reader Shld Riser Rated**

Card Reader (Shielded), Lock Power (N/S), Request Exit (N/S), and Door Contact (N/S) components; each component jacketed; 4 components cabled and bound together.

Card Reader: 32SR.223* style - 3 pair 22 AWG overall shielded* and jacketed component. Identified as "Card Reader" – the means of such inner component identification at cable manufacturer's option. (*SDN 36MRS substitutes 31SR.226 6C 22 Shielded.)

Lock Power: 31UR.184 style - 4 conductor 18 AWG unshielded, jacketed component. Identified as "Lock Power" – the means of such inner component identification at cable manufacturer's option.

Request Exit: 31UR.224 style - 4 conductor 22 AWG unshielded jacketed component. Identified as "Request Exit" and/or "REX" – the means of such inner component identification at cable manufacturer's option.

Door Contact: 31UR.222 style - 2 conductor 22 AWG unshielded jacketed component. Identified as "Door Contact" – the means of such inner component identification at cable manufacturer's option.

Assembly: 4 jacketed components cabled and held together by physical means as a contiguous unit.

Shielding: No overall shielding. Individual components shielded as described above.

Binding: 4 jacketed components will be bound together by a physical means, not merely by inertia. This may be by use of an overall jacket, or an extruded inner spline adhering to all four components, or periodically spaced kisses adhering to all for components, or outer binder string or strings wrapped with a long lay length.

Jacket: Where binding is accomplished with an overall jacket it shall be PVC jacket of thickness according to UL 444.

Description and treatment: A jacketed assembly is defined as a single cable, and the design rules, installation practices, and installation time called for will be as for a single cable. Where binding is accomplished with an inner spline, or periodic kisses, or binder string(s), the description and treatment of the assembly is as of 4 separate cables. The design rules, installation practices, and installation time called for shall be as for four cables. Cables without an outer jacket shall not be pulled during installation.

Diameter: Nominal diameter shall not exceed that in table 36MR Column B.

Flame test: Jacketed styles (cables) shall pass UL 1666, for use in risers. Other styles (assemblies) – each component shall pass UL 1666 individually, as the individual and separate cables they are. Such testing to have been performed by Underwriters Laboratories (UL), or their designate. CMR product listing shall be maintained by UL.

Voltage: 300V.

Application: Communication Cable per NEC Art. 800 and UL 444; also for Class 3 or Class 2 Remote-Control, Signaling, and Power-Limited Circuits per NEC Art. 725.

Installation: Suitable for open cabling use in Risers as CMR, CL3R, or CL2R; also for open cabling horizontal runs since rated higher than CM, CL3, CL2 respectively. Not for use in plenum spaces unless running in conduit.

36MR: 300V Specification 36: Access Control Door Composite Cable Card Reader Shld Riser Rated

Card Reader (Shielded), Lock Power (N/S), Request Exit (N/S), and Door Contact (N/S) components; each component jacketed; 4 components cabled and bound together.

Continued

Temp.: 0C to 60C minimum range, continuous operating ambient temperature.

QA: UL Listed as cable type CMR when a specified a single jacketed cable. When specified as an assembly of four cables (spline, kisses, or binder strings for binding) – the individual cables shall be UL Listed as cable type CMR. Cable ratings shall be printed on the jacket(s).
UL file number will be printed on the Jacket(s).
UL label shall be displayed on every cable package near manufacturer's label.
Assemblies of four cables shall display four UL labels on their shared reel.

Packaging: 1000ft or 500ft: spools or reels.

Appr. Manf: Manufacturer make and model shall be as listed in table 36MR.

Availability: Cable shall be available from local distribution known to project management.

Table 36MR Access Control Door Composite Cable Card Reader Shld Riser Rated

Const	SDN	B	C	TAPPAN	GENERAL CABLE	Belden West Penn
31UR.184 + 32SR.223 + 31UR.222 + 31UR.224 Jacketed Cable CMR	36MRJ	.775	125	H92091	4ERS1S	-----
31UR.184 + 31SR.226* + 31UR.222 + 31UR.224 4 CMR rated Cables; An assembly	36MRS*	.440	115	-----	-----	LS1822*

* Type 36MRS substitutes for card reader: 6/C shielded in place of 3 pr OAS.

Manufacturer's makes vary. Diameters (Col. B) and weights (Col. C) are conservative and for engineering purposes only. Consult manufacturer's data for individual make & model information.

Column B: Nominal Cable Diameter in inches, for conduit fill calculations.
Column C: Nominal cable weight in pounds per 1000ft.

36MP: 300V Specification 36: Access Control Door Composite Cable **Card Reader Shld Plenum Rated**

Card Reader (Shielded), Lock Power (N/S), Request Exit (N/S), and Door Contact (N/S) components; each component jacketed; 4 components cabled and bound together.

Card Reader: 32SP*.223 style - 3 pair 22 AWG overall shielded* and jacketed component. Identified as "Card Reader" – the means of such inner component identification at cable manufacturer's option.(*SDN 36MPS substitutes 31SP.226 6C 22 AWG shielded.)

Lock Power: 31UP.184 style - 4 conductor 18 AWG unshielded, jacketed component. Identified as "Lock Power" – the means of such inner component identification at cable manufacturer's option.

Request Exit: 31UP.224 style - 4 conductor 22 AWG unshielded jacketed component. Identified as "Request Exit" and/or "REX" – the means of such inner component identification at cable manufacturer's option.

Door Contact: 31UP.222 style - 2 conductor 22 AWG unshielded jacketed component. Identified as "Door Contact" – the means of such inner component identification at cable manufacturer's option.

Assembly: 4 jacketed components cabled and held together by physical means as a contiguous unit.

Shielding: No overall shielding. Individual components shielded as described above.

Binding: 4 jacketed components will be bound together by a physical means, not merely by inertia. This may be by use of an overall jacket, or an extruded inner spline adhering to all four components, or periodically spaced kisses adhering to all for components, or outer binder string or strings s wrapped with a long lay length.

Jacket: Where binding is accomplished with an overall jacket it shall be LS-PVC or PVDF jacket of thickness according to UL 444.

Description and treatment: A jacketed assembly is defined as a single cable, and the design rules, installation practices, and installation time called for will be as for a single cable. Where binding is accomplished with an inner spline, or periodic kisses, or binder string(s), the description and treatment of the assembly is as of 4 separate cables. The design rules, installation practices, and installation time called for shall be as for four cables. Cables without an outer jacket shall not be pulled during installation.

Diameter: Nominal diameter shall not exceed that in table 36MP Column B.

Flame test: Jacketed styles (cables) shall pass NFPA 262 for use in plenum spaces. Other styles (assemblies) – each component shall pass NFPA 262 individually, as the individual and separate cables they are. Such testing to have been performed by Underwriters Laboratories (UL), or their designate. CMP product listing shall be maintained by UL.

Voltage: 300V.

Application: Communication Cable per NEC Art. 800 and UL 444; also for Class 3 or Class 2 Remote-Control, Signaling, and Power-Limited Circuits per NEC Art. 725.

Installation: Suitable for open cabling use in Plenums as CMP. Also for open cabling horizontal runs or running in risers, since rated higher than CM, and CMR respectively.

36MP: 300V Specification 36: Access Control Door Composite Cable **Card Reader Shld Plenum Rated**

Card Reader (Shielded), Lock Power (N/S), Request Exit (N/S), and Door Contact (N/S) components; each component jacketed; 4 components cabled and bound together.

Continued

Temp.: 0C to 60C minimum range, continuous operating ambient temperature.

QA: UL Listed as cable type CMP when a specified a single jacketed cable. When specified as an assembly of four cables (spline, kisses, or binder strings for binding) – the individual cables shall be UL Listed as cable type CMP. Cable ratings shall be printed on the jacket(s).
UL file number will be printed on the jacket(s).
UL label shall be displayed on every cable package near manufacturer's label.
Assemblies of four cables shall display four UL labels on their shared reel.

Packaging: 1000ft or 500ft: spools or reels.

Appr. Manf: Manufacturer make and model shall be as listed in table 36MP.

Availability: Cable shall be available from local distribution known to project management.

Table 36MP Access Control Door Composite Cable Card Reader Shld Plenum Rated

Const	SDN	B	C	TAPPAN	GENERAL CABLE	BELDEN-WEST PENN
31UP.184 + 32SP.223 + 31UP.222 + 31UP.224 Jacketed Cable CMP	36MPJ	.495	120	H91602	4EPL1S	AC251822B3P
31UP.184 + 31SP.226* + 31UP.222 + 31UP.224 4 CMP rated Cables; An assembly	36MPS*	.425	110	-----	-----	LS251822

* Type 36MPS substitutes for card reader: 6/C shielded in place of 3 pr OAS.

Manufacturer's makes vary. Diameters (Col. B) and weights (Col. C) are conservative and for engineering purposes only. Consult manufacturer's data for individual make & model information.

Column B: Nominal Cable Diameter in inches, for conduit fill calculations.
Column C: Nominal cable weight in pounds per 1000ft.

36SR: 300V Specification 36: Access Control Door Composite Cable **All Legs Shielded Riser Rated**

Card Reader (Shld), Lock Power (Shld), Request Exit (Shld), and Door (Contact Shld); each component shielded (S) and jacketed; 4 components cabled and bound together.

Card Reader: 32SR.223 style - 3 pair 22 AWG overall shielded and jacketed component. Identified as "Card Reader" – the means of such inner component identification at cable manufacturer's option.

Lock Power: 31SR.184 style - 4 conductor 18 AWG unshielded, jacketed component. Identified as "Lock Power" – the means of such inner component identification at cable manufacturer's option.

Request Exit: 31SR.224 style - 4 conductor 22 AWG unshielded jacketed component. Identified as "Request Exit" and/or "REX" – the means of such inner component identification at cable manufacturer's option.

Door Contact: 31SR.222 style - 2 conductor 22 AWG unshielded jacketed component. Identified as "Door Contact" – the means of such inner component identification at cable manufacturer's option.

Assembly: 4 jacketed components cabled and held together by physical means as a contiguous unit.

Shielding: No overall shielding. Individual components shielded as described above.

Binding: 4 jacketed components will be bound together by a physical means, not merely by inertia. This may be by use of an overall jacket, or an extruded inner spline adhering to all four components, or periodically spaced kisses adhering to all for components, or outer binder string or strings wrapped with a long lay length.

Jacket: Where binding is accomplished with an overall jacket it shall be PVC jacket of thickness according to UL 444.

Description and treatment: A jacketed assembly is defined as a single cable, and the design rules, installation practices, and installation time called for will be as for a single cable. Where assembly is accomplished with an inner spline, or periodic kisses, or binder string(s), the description and treatment of the assembly is as of 4 separate cables. The design rules, installation practices, and installation time called for shall be as for four cables. Cables without an outer jacket shall not be pulled during installation.

Diameter: Nominal diameter shall not exceed that in table 36SR Column B.

Flame test: Jacketed styles (cables) shall pass UL 1666, for use in risers. Other styles (assemblies) – each component shall pass UL 1666 individually, as the individual and separate cables they are. Such testing to have been performed by Underwriters Laboratories (UL), or their designate. CMR product listing shall be maintained by UL.

Voltage: 300V.

Application: Communication Cable per NEC Art. 800 and UL 444; also for Class 3 or Class 2 Remote-Control, Signaling, and Power-Limited Circuits per NEC Art. 725.

Installation: Suitable for open cabling use in Risers as CMR, CL3R, or CL2R; also for open cabling horizontal runs since rated higher than CM, CL3, CL2 respectively. Not for use in plenum spaces unless running in conduit.

36SR: 300V Specification 36: Access Control Door Composite Cable All Legs Shielded Riser Rated

Card Reader (Shld), Lock Power (Shld), Request Exit (Shld), and Door Contact (Shld); each shielded (S) and jacketed; 4 components cabled and bound together.

Continued

Temp.: 0C to 60C minimum range, continuous operating ambient temperature.

QA: UL Listed as cable type CMR when a specified a single jacketed cable. When specified as an assembly of four cables (spline, kisses, or binder strings for binding) – the individual cables shall be UL Listed as cable type CMR. Cable ratings shall be printed on the jacket(s).
UL file number will be printed on the jacket(s).
UL label shall be displayed on every cable package near manufacturer's label.
Assemblies of four cables shall display four UL labels on their shared reel.

Packaging: 1000ft or 500ft: spools or reels.

Appr. Manf: Manufacturer make and model shall be as listed in table 36SR.

Availability: Cable shall be available from local distribution known to project management.

Table 36SR Access Control Door Composite Cable All Legs Shld Riser Rated

Const	SDN	B	C	TAPPAN	GENERAL CABLE	BELDEN
31SR.184 + 32SR.223 + 31SR.222 + 31SR.224 Jacketed Cable CMR	36SRJ	.565	130	H91765	4ERS4S	558AFJ
31SR.184 + 32SR.223 + 31SR.222 + 31RR.224 4 CMR rated Cables; An assembly	36SRS	.448	105	-----	-----	558AFS

Manufacturer's makes vary. Diameters (Col. B) and weights (Col. C) are conservative and for engineering purposes only. Consult manufacturer's data for individual make & model information.

Column B: Nominal Cable Diameter in inches, for conduit fill calculations.
Column C: Nominal cable weight in pounds per 1000ft.

36SP: 300V Specification 36: Access Control Door Composite Cable **All Legs Shielded Plenum Rated**

Card Reader Shld, Lock Power Shld, Request Exit (REX) Shld, and Door Contact Shld; each shielded (S) and jacketed; 4 jacketed cables cabled and bound together.

Card Reader: 32SP.223 style - 3 pair 22 AWG overall shielded and jacketed component. Identified as "Card Reader" – the means of such inner component identification at cable manufacturer's option.

Lock Power: 31SP.184 style - 4 conductor 18 AWG unshielded, jacketed component. Identified as "Lock Power" – the means of such inner component identification at cable manufacturer's option.

Request Exit: 31SP.224 style - 4 conductor 22 AWG unshielded jacketed component. Identified as "Request Exit" and/or "REX" – the means of such inner component identification at cable manufacturer's option.

Door Contact: 31SP.222 style - 2 conductor 22 AWG unshielded jacketed component. Identified as "Door Contact" – the means of such inner component identification at cable manufacturer's option.

Assembly: 4 jacketed components cabled and held together by physical means as a contiguous unit.

Shielding: No overall shielding. Individual components shielded as described above.

Binding: 4 jacketed components will be bound together by a physical means, not merely by inertia. This may be by use of an overall jacket, or an extruded inner spline adhering to all four components, or periodically spaced kisses adhering to all for components, or outer binder string or strings wrapped with a long lay length.

Jacket: Where binding is accomplished with an overall jacket it shall be LS-PVC or PVDF jacket of thickness according to UL 444.

Description and treatment: A jacketed assembly is defined as a single cable, and the design rules, installation practices, and installation time called for will be as for a single cable. Where binding is accomplished with an inner spline, or periodic kisses, or binder string(s), the description and treatment of the assembly is as of 4 separate cables. The design rules, installation practices, and installation time called for shall be as for four cables. Cables without an outer jacket shall not be pulled during installation.

Diameter: Nominal diameter shall not exceed that in table 36SP Column B.

Flame test: Jacketed styles (cables) shall pass NFPA 262 for use in plenum spaces. Other styles (assemblies) – each component shall pass NFPA 262 individually, as the individual and separate cables they are. Such testing to have been performed by Underwriters Laboratories (UL), or their designate. CMP product listing shall be maintained by UL.

Voltage: 300V.

Application: Communication Cable per NEC Art. 800 and UL 444; also for Class 3 or Class 2 Remote-Control, Signaling, and Power-Limited Circuits per NEC Art. 725.

Installation: Suitable for open cabling use in Plenums as CMP. Also for open cabling horizontal runs or running in risers, since rated higher than CM, and CMR respectively.

36SP: 300V Specification 36: Access Control Door Composite Cable **All Legs Shielded Plenum Rated**

Card Reader Shld, Lock Power Shld, Request Exit (REX) Shld, and Door Contact Shld; each shielded (S) and jacketed; 4 components cabled and bound together.

Continued

Temp.: 0C to 60C minimum range, continuous operating ambient temperature.

QA: UL Listed as cable type CMP when a specified a single jacketed cable. When specified as an assembly of four cables (spline, kisses, or binder strings for binding) – the individual cables shall be UL Listed as cable type CMP. Cable ratings shall be printed on the jacket(s).
UL file number will be printed on the Jacket(s).
UL label shall be displayed on every cable package near manufacturer's label.
Assemblies of four cables shall display four UL labels on their shared reel.

Packaging: 1000ft or 500ft: spools or reels.

Appr. Manf: Manufacturer make and model shall be as listed in table 36SP.

Availability: Cable shall be available from local distribution known to project management.

Table 36SP Access Control Door Composite Cable All Legs Shld Plenum Rated

Const	SDN	B	C	TAPPAN	GENERAL CABLE	BELDEN
31SP.184 + 32SP.223 + 31SP.222 + 31SP.224 Jacketed Cable CMP	36SPJ	.492	120	H91601	4EPL4S	658AFJ
31SP.184 + 32SP.223 + 31SP.222 + 31SP.224 4 CMP rated Cables; Made as an assembly	36SPS	.420	100	-----	-----	658AFS

Manufacturer's makes vary. Diameters (Col. B) and weights (Col. C) are conservative and for engineering purposes only. Consult manufacturer's data for individual make & model information.

Column B: Nominal Cable Diameter in inches, for conduit fill calculations.
Column C: Nominal cable weight in pounds per 1000ft.

27UH: 150V Specification 27: Highly Stranded Bare Copper; Cabled **Unshielded**
High count Str. BC / PVC or PP Insulation / Cabled / Unshielded (U) / PVC Jacket **Horizontal Rated.**

Speaker Cable for Permanent Concealed Installation

Conductor: Stranded Bare Copper. Minimum number of strands per Table 27UH Column A. Consult individual manufacturers' specifications.

Insulation: Polypropylene (preferred) or PVC, minimum thickness according to UL 13.

Color Code: Black, Red, White, Green. <> 2C Cables, Black. White is permitted.

Assembly: Twisted together so that a 2 conductor cable is also considered 1 pair and 4 conductor cable is considered 1 quad. Ripcord under the jacket to facilitate removal.

Shielding: None.

Jacket: PVC. Minimum thickness according to UL 13 for CL2 cable.

Diameter: Nominal diameter shall not exceed that in table 27UH Column B.

Flame test: As a minimum, passes UL 1685, for general purpose use.

Voltage: 150V NEC CL2, minimum.

Application: 10 AWG is expressly FOR AUDIO USE ONLY.
Class 2 Remote-Control, Signaling, and Power-Limited Circuits per NEC Art. 725 & UL 13.
CL3 rating is optional. CM rating is optional where applicable.

Installation: Suitable for open cabling use in horizontal runs and general purpose use. Not for use in risers unless rated CL2R, CL3R, or CMR. Not for use in plenum spaces unless running in conduit.

Temp.: 0C to 60C minimum range, continuous operating ambient temperature.

QA: UL Listed as cable type CL2 or higher. Cable rating(s) shall be printed on the jacket.
UL file number will be printed on the jacket alongside the gauge size and cable rating.
UL label shall be displayed on every cable package near manufacturer's product label.

Packaging: 1000ft or 500ft: spools, reels or boxes.
Appr. Manf: Manufacturer make and model shall be as listed in table 27UH.
Availability: Cable shall be available from local distribution known to project management.

Table 27UH High Str. BC; PP or PVC Insul.; Cabled; Unshielded; Horizontal Rated

Const	SDN	A	B	C	TAPPAN	GENERAL CABLE	BELDEN-WEST PENN	BELDEN 13--A	BELDEN 5---UP
16/2	27UH.162	65	.222	28	A50322	162SSA65	HA225	1307A	5200UP
16/4	27UH.164	65	.270	52	A50323	164SSA65	-----	1308A	5202UP
14/2	27UH.142	41	.264	44	A60155	142SSA41	HA226	1309A	5100UP
14/4	27UH.144	41	.319	82	A60156	144SSA41	-----	1310A	5102UP
12/2	27UH.122	65	.352	70	A70042	122SSA105	HA227	1311A	5000UP
12/4	27UH.124	65	.423	125	A70043	124SSA105	-----	1312A	5002UP
10/2	27UH.102	65	.428	104	A80001	-----	HA210	1313A	5T00UP
10/4	27UH.104	105	.400	186	A80002	-----	-----	-----	-----

Manufacturer's makes vary. Diameters (Col. B) and weights (Col. C) are conservative and for engineering purposes only. Consult manufacturer's data for individual make & model information.

Column A: Minimum number of strands. Consult manufacturer's spec.'s for higher strands.
Column B: Cable Diameter in inches. Manufacturers' vary. Consult manufacturer's spec.'s.
Column C: Nominal cable weight in pounds per 1000ft

27UP: 150V Specification 27: Stranded Bare Copper; Cabled **Unshielded**
Str BC / LS-PVC Insulation / Cabled / Unshielded (U) / LS-PVC Jacket. **Plenum Rated (P).**

Speaker Cable for Permanent Installation

Conductor: Stranded Bare Copper. Minimum number of strands per Table 27UP Column A.

Insulation: LS-PVC or fluoropolymer, minimum thickness according to UL 13, minimum.

Color Code: Black, Red, White, Green. <> Black, White for 2/C cables is permitted.

Assembly: Twisted together so that a 2 conductor cable is also considered 1 pair and 4 conductor cable is considered 1 quad. Ripcord under the jacket to facilitate removal.

Shielding: None.

Jacket: LS-PVC. Minimum thickness according to UL 13.

Diameter: Nominal diameter shall not exceed that in table 27UP Column B.

Flame test: Passes NFPA 262 (UL 910 Steiner Tunnel) flame test for use in plenum spaces.

Voltage: 150V (minimum) rating according to manufacturer's specification.

Application: 10 AWG is expressly FOR AUDIO USE ONLY.
Class 2 Remote-Control, Signaling, and Power-Limited Circuits per NEC Art. 725 & UL 13.
CL3 rating is optional. CM rating is optional where applicable.

Installation: Suitable for open cabling use in plenum spaces as CL2P Also for open cabling use in risers and in horizontal runs since rated higher than those requirements..

Temp.: 0C to 60C minimum range, continuous operating ambient temperature.

QA: UL Listed as cable type CL2P or higher. Cable rating(s) shall be printed on the jacket. UL file number will be printed on the jacket alongside the gauge size and cable rating. UL label shall be displayed on every cable package near manufacturer's product identification label.

Packaging: 1000ft or 500ft: spools, reels or boxes.

Appr. Manf: Manufacturer make and model shall be as listed in table 27UP.

Availability: Cable shall be available from local distribution known to project management.

Table 27UP **Str. BC; LS-PVC Insul.; Cabled; N/S; Plenum Rated**

Const	SDN	A	B	C	TAPPAN	GENERAL CABLE	BELDEN-WEST PENN	BELDEN
16/2	27UP.162	19	.178	26	P50032	E3042S	25225B	6200UE
16/4	27UP.164	19	.221	52	P50059	E3044S	25245B	6202UE
14/2	27UP.142	19	.216	38	P60040	E3052S	25226B	6100UE
14/4	27UP.144	19	.265	73	P60013	E3054S	25246B	6102UE
12/2	27UP.122	19	.264	56	P70031	E3062S	25227B	6000UE
12/4	27UP.124	19	.314	111	P70019	E3064S	25248B	6002UE
10/2	27UP.102	19	.324	88	P80004	-----	25210	6T00UP
10/4	27UP.104	19	.390	168	P80002	-----	-----	-----

Manufacturer's makes vary. Diameters (Col. B) and weights (Col. C) are conservative and for engineering purposes only. Consult manufacturer's data for individual make & model information.

Column A: Minimum number of strands.
Column B: Cable Diameter in inches. Manufacturers' vary. Consult manufacturer's spec.'s.
Column C: Nominal cable weight in pounds per 1000ft

MASTER TABLE OF PUBLIC CABLE 1.0 PRODUCTS

PERIODIC TABLE OF PARTIAL CONTENTS – Detail for Solid Bare Copper, UL minimum Insulation

Cabled; Solid Bare Copper; UL Minimum Insulation; **Specification 311**

Shielding	Installation	Spec / Table	Spec. Page	Table Page
Unshielded	Riser	311UR	68	68
	Plenum	311UP	69	69
Shielded	Riser	311SR	70	70
	Plenum	311SP	71	71

Multiple Unshielded Pairs; Solid Bare Copper; UL Minimum Insulation; **Specification 312**

Shielding	Installation	Spec / Table	Spec. Page	Table Page
Unshld. Pairs / No OAS	Riser	312UR	72	72
	Plenum	312UP	73	73
Unshld. Pairs / OAS	Riser	312SR	74	74
	Plenum	312SP	75	75

Multiple Pairs; EPS; Solid Bare Copper; UL Minimum Insulation; **Specification 314**

Shielding	Installation	Table	Spec. Page	Table Page
EPS	Riser	314UR	76	76
	Plenum	314UP	77	77

Combo Cables; Solid BC; UL Minimum Insul.; Mixed Shielding /AWG Sizes **Specification 315**

Shielding	Installation	Table	Spec. Page	Table Page
Mixed "In & Out"	Riser	315MR	78	78
	Plenum	315MP	79	79

PERIODIC TABLE OF PARTIAL CONTENTS – Detail for Stranded TC, Upgrade Insulation (partial)

Cabled; Stranded Tinned Copper; Upgrade Insulation; **Specification 321**

Shielding	Installation	Spec / Table	Spec. Page	Table Page
Unshielded	Horizontal	321UH	82	83
	Plenum	321UP	84	85
Shielded	Horizontal	321SH	86	87
	Plenum	321SP	88	89 – 90

Multiple Unshielded Pairs; Stranded Tinned Copper; Upgrade Insulation; **Specification 322**

Shielding	Installation	Spec / Table	Spec. Page	Table Page
Unshld. Pairs / No OAS	Horizontal	322UH	92	93
	Plenum	322UP	94	95
Unshld. Pairs / OAS	Horizontal	322SH	96	97
	Plenum	322SP	98	99

Multiple Pairs; EPS/CD; Stranded Tinned Copper; Upgrade Insulation; **Specification 323**

Shielding	Installation	Table	Spec. Page	Table Page
EPS/CD	Horizontal	323UH	100	102
	Plenum	323UP	104	105

Multiple Pairs; EPS; Stranded Tinned Copper; Upgrade Insulation; **Specification 324**

Shielding	Installation	Table	Spec. Page	Table Page
EPS	Horizontal	324UH	101	102 – 103
	Plenum	324UP	106	107

311UR: 300V Specification 311: Solid Bare Copper; Cabled; Unshielded
Solid BC / PVC or PP Insulation / Cabled / Unshielded (U) / PVC Jacket. Riser Rated (R).

Conductor: Solid Bare Copper.

Insulation: PVC or Polypropylene, thickness according to UL444 or UL 13.

Color Code: Black, Red, White, Green, Brown, Blue, Orange, Yellow, Violet, Gray, Pink, Beige.

Assembly: Twisted together. 2C is 1 pair, 3C is 1 Triad, and 4C is one quad.
Ripcord under the jacket to facilitate removal.

Shielding: None.

Jacket: PVC. Thickness according to UL 444 or UL 13 as applicable.
Color: Gray unless specified.

Diameter: Nominal diameter shall not exceed that in table 311UR Column B.

Flame test: Passes UL 1666, for use in risers.

Voltage: 300V but not printed on the jacket per UL 444 and UL 13.

Application: Communication Cable per NEC Art. 800 and UL 444 and Class 3 or Class 2
Remote-Control, Signaling, and Power-Limited Circuits per NEC Art. 725 and UL 13.

Installation: Suitable for open cabling use in Risers as CMR, CL3R, or CL2R; also for open cabling in horizontal runs since rated higher than CM, CL3, CL2 respectively.
Not for use in plenum spaces unless running in conduit.

Temp.: 0C to 60C minimum range, continuous operating ambient temperature.

QA: UL Listed as cable type CMR or CL3R, or both. Cable rating(s) shall be printed on the jacket.
UL file number will be printed on the jacket alongside the gauge size and cable rating.
UL label shall be displayed on every cable package near manufacturer's product label.

Packaging: 1000ft or 500ft: spools, reels or boxes.

Appr. Manf: Manufacturer make and model shall be as listed in table 311UR.

Availability: Cable shall be available from local distribution known to project management

Table 311UR Solid BC; PVC or PP Insul.; Cabled; Unshielded; Riser Rated

Const	SDN	B	C	TAPPAN	GENERAL CABLE	BELDEN-WEST PENN	BELDEN
22/2	311UR.222	.122	9	R20374	E1000S	220	5520UE
22/3	311UR.223	.141	12	R20375	-----	230	-----
22/4	311UR.224	.142	16	R20376	E1001S	240	5522UE
22/6	311UR.226	.168	22	R20393	-----	-----	5524UE
22/8	311UR.228	.182	28	R20987	-----	-----	5526UE
22/12	311UR.2212	.221	42	R20381	-----	-----	5529UE
18/2	311UR.182	.157	17	R40510	E1030S	-----	5320UE
18/3	311UR.183	.167	23	R40511	-----	-----	5321UE
18/4	311UR.184	.195	30	R40512	-----	-----	5322UE

Manufacturer's makes vary. Diameters (Col. B) and weights (Col. C) are conservative and for engineering purposes only. Consult manufacturer's data for individual make & model information.

Column B: Nominal Cable Diameter in inches, for conduit fill calculations.
Column C: Nominal cable weight in pounds per 1000ft.

311UP: 300V Specification 311: Solid Bare Copper; Cabled; Unshielded
Solid BC / LS-PVC Insulation / Cabled / Unshielded (U) / LS-PVC Jacket. Plenum Rated (P).

Conductor: Solid Bare Copper.

Insulation: LS-PVC, thickness according to UL 444 or UL 13 as applicable.

Color Code: Black, White, Red, Green, Brown, Blue

Assembly: Twisted together. 2C is 1 pair, 3C is 1 Triad, and 4C is one quad.
Ripcord under the jacket to facilitate removal.

Shielding: None.

Jacket: LS-PVC. Thickness according to UL 444 or UL 13 as applicable.
Color: Natural or White, or gray; unless specified.

Diameter: Nominal diameter shall not exceed that in table 311UP Column B.

Flame test: Passes NFPA 262 (UL 910 Steiner Tunnel) flame test for use in plenum spaces.

Voltage: 300V but not printed on the jacket per UL 444 and UL 13.

Application: Communication Cable per NEC Art. 800 and UL 444; or Class 3 or Class 2 Remote-Control, Signaling, and Power-Limited Circuits per NEC Art. 725 and UL 13.

Installation: Suitable for open cabling use in plenum spaces as CMP, CL3P or CL2P; also for open cabling use in risers and in horizontal runs since rated higher than those requirements.

Temp.: 0C to 60C minimum range, continuous operating ambient temperature.

QA: UL Listed as cable type CMP, CL3P, or both. Cable rating(s) shall be printed on the jacket. UL file number will be printed on the jacket alongside the gauge size and cable rating. UL label shall be displayed on every cable package near manufacturer's product identification label.

Packaging: 1000ft or 500ft: spools, reels or boxes.

Appr. Manf: Manufacturer make and model shall be as listed in table 311UP.

Availability: Cable shall be available from local distribution known to project management

Table 311UP Solid BC; LS-PVC Insul.; Cabled; Unshielded; Plenum Rated

Const	SDN	B	C	TAPPAN	GENERAL CABLE	BELDEN-WEST PENN	BELDEN
22/2	311UP.222	.118	9	P20175	E3000S	-----	6520UE
22/3	311UP.223	.128	13	P20170	-----	-----	6521UE
22/4	311UP.224	.136	16	P20177	E3001S	-----	6522UE
22/6	311UP.226	.162	23	P20628	-----	-----	6524UE
18/2	311UP.182	.147	17	P40559	E3030S	-----	6320UE
18/3	311UP.183	.156	24	P40589	-----	-----	6321UE
18/4	311UP.184	.183	30	P40555	-----	-----	6322UE

Manufacturer's makes vary. Diameters (Col. B) and weights (Col. C) are conservative and for engineering purposes only. Consult manufacturer's data for individual make & model information.

Column B: Nominal Cable Diameter in inches, for conduit fill calculations.
Column C: Nominal cable weight in pounds per 1000ft.

311SR: 300V Specification 311: Solid Bare Copper; Cabled **Shielded**
Solid BC / PVC or PP Insulation / Cabled / Shielded (S) / PVC Jacket. **Riser Rated (R).**

Conductor: Solid Bare Copper.

Insulation: PVC, thickness according to UL 444 or UL 13 as applicable.

Color Code: Black, Red, White, Green, Brown, Blue, Orange, Yellow, Violet, Gray, Pink, Beige.

Assembly: Twisted together. 2C is 1 pair, 3C is 1 Triad, and 4C is one quad.
Ripcord under the jacket to facilitate removal.

Shielding: Aluminum Mylar Shielding Tape, Tinned Copper Drain Wire. Aluminum foil surface facing out; drain wire is outside the tape and in continuous contact with the foil.

Jacket: PVC. Thickness according to UL 444 or UL 13 as applicable. Color: Gray unless specified.

Diameter: Nominal diameter shall not exceed that in table 311SR Column B.

Flame test: Passes UL 1666, for use in risers.

Voltage: 300V but not printed on the jacket per UL 444 and UL 13.

Application: Communication Cable per NEC Art. 800 and UL 444; or Class 3 or Class 2 Remote-Control, Signaling, and Power-Limited Circuits per NEC Art. 725 and UL 13.

Installation: Suitable for open cabling use in Risers as CMR, CL3R, or CL2R; also for open cabling in horizontal runs since rated higher than CM, CL3, CL2 respectively.
Not for use in plenum spaces unless running in conduit.

Temp.: 0C to 60C minimum range, continuous operating ambient temperature.

QA: UL Listed as cable type CMR or CL3R, or both. Cable rating shall be printed on the jacket.
UL file number will be printed on the jacket alongside the gauge size and cable rating.
UL label shall be displayed on every cable package near manufacturer's product label.

Packaging: 1000ft or 500ft: spools, reels or boxes.
Appr. Manf: Manufacturer make and model shall be as listed in table 311SR.
Availability: Cable shall be available from local distribution known to project management

Table 311SR **Solid BC; PVC or PP Insul.; Cabled; Shielded; Riser Rated**

Const	SDN	B	C	TAPPAN	GENERAL CABLE	BELDEN-WEST PENN	BELDEN
22/2	311SR.222	.125	14	R20540	E2000S	290	5520FE
22/3	311SR.223	.134	16	R20383	-----	300	5521FE
22/4	311SR.224	.140	17	R20384	-----	-----	-----
22/6	311SR.226	.156	22	R20386	-----	-----	-----
22/8	311SR.228	.178	28	R20387	-----	-----	-----
22/12	311SR.2212	.204	39	R20389	-----	-----	-----
20/2	311SR.202	.136	14	R30178	-----	-----	-----
20/3	311SR.203	.153	20	R30199	-----	-----	5421FE
18/2	311SR.182	.157	21	R40518	E2030S	-----	5320FE
18/3	311SR.183	.167	26	R40519	-----	-----	-----
18/4	311SR.184	.195	33	R40520	-----	-----	-----

Manufacturer's makes vary. Diameters (Col. B) and weights (Col. C) are conservative and for engineering purposes only. Consult manufacturer's data for individual make & model information.

Column B: Nominal Cable Diameter in inches, for conduit fill calculations.
Column C: Nominal cable weight in pounds per 1000ft.

311SP: 300V Specification 311: Solid Bare Copper; Cabled Shielded
Solid BC / LS-PVC Insulation / Cabled / Shielded (S) / LS-PVC Jacket. Plenum Rated (P).

Conductor: Solid Bare Copper.

Insulation: LS-PVC, thickness according to UL 444 or UL 13 as applicable.

Color Code: Black, Red, White, Green, Brown, Blue, Orange, Yellow, Violet, Gray, Pink, Beige.

Assembly: Twisted together. 2C is 1 pair, 3C is 1 Triad, and 4C is one quad.
Ripcord under the jacket to facilitate removal.

Shielding: Aluminum Mylar Shielding Tape, Tinned Copper Drain Wire. Aluminum foil surface facing out; drain wire is outside the tape and in continuous contact with the foil.

Jacket: LS-PVC. Thickness according to UL 444 or UL 13 as applicable. Color: Gray unless specified.

Diameter: Nominal diameter shall not exceed that in table 311SP Column B.

Flame test: Passes NFPA 262 (UL 910 Steiner Tunnel) flame test for use in plenum spaces.

Voltage: 300V but not printed on the jacket per UL 444 and UL 13.

Application: Communication Cable per NEC Art. 800 and UL 444; or Class 3 or Class 2 Remote-Control, Signaling, and Power-Limited Circuits per NEC Art. 725 and UL 13.

Installation: Suitable for open cabling use in plenum spaces as CMP, CL3P or CL2P; also for open cabling use in risers and in horizontal runs since rated higher than those requirements..

Temp.: 0C to 60C minimum range, continuous operating ambient temperature.

QA: UL Listed as cable type CMP, CL3P, or both. Cable rating shall be printed on the jacket.
UL file number will be printed on the jacket alongside the gauge size and cable rating.
UL label shall be displayed on every cable package near manufacturer's product label.

Packaging: 1000ft or 500ft: spools, reels or boxes.

Appr. Manf: Manufacturer make and model shall be as listed in table 311SP.

Availability: Cable shall be available from local distribution known to project management

Table 311SP Solid BC; PVC or PP Insul.; Cabled; Shielded; Plenum Rated

Const	SDN	B	C	TAPPAN	GENERAL CABLE	BELDEN-WEST PENN	BELDEN
22/2	311SP.222	.122	13	P20178	E2100S	25290B	6520FE
22/3	311SP.223	.133	16	P20425	-----	25300B	6521FE
22/4	311SP.224	.140	17	P20180	-----	-----	-----
22/6	311SP.226	.146	22	P20305	-----	-----	-----
22/8	311SP.228	.166	28	P20308	-----	-----	-----
22/12	311SP.2212	.192	39	P20641	-----	-----	-----
20/2	311SP.202	.134	15	P30115	-----	-----	6420FE
18/2	311SP.182	.150	22	P40574	E2200S	-----	6320FE
18/3	311SP.183	.155	25	P40557	-----	-----	-----
18/4	311SP.184	.183	32	P40558	-----	-----	-----

Manufacturer's makes vary. Diameters (Col. B) and weights (Col. C) are conservative and for engineering purposes only. Consult manufacturer's data for individual make & model information.

Column B: Nominal Cable Diameter in inches, for conduit fill calculations.
Column C: Nominal cable weight in pounds per 1000ft.

312UR: 300V Specification 312: Solid Bare Copper; Pairs (N/S) Unshielded O/A
Solid BC / PVC or PP Insul. / Multiple Pairs (N/S) / Unshielded (U) / PVC Jacket. Riser Rated (R).

Conductor: Solid Bare Copper.

Insulation: PVC or Polypropylene, thickness according to UL 444.

Color Code: Pair (1) Black& Red, (2) Black&White, (3) Black&Green, (4) Black&Blue, (5) Black&Yellow, (6) Black&Brown, (7) Black&Orange, (8) Red&White, (9) Red&Green, (10) Red&Blue, (11) Red&Yellow, (12)Red&Brown. Preferred.

Assembly: Insulated conductors twisted into pairs. Multiple pairs cabled together. Ripcord under the jacket to facilitate removal.

Shielding: None.

Jacket: PVC. Thickness according to UL 444. Color: Gray unless specified.

Diameter: Nominal diameter shall not exceed that in table 312UR Column B.

Flame test: Passes UL 1666, for use in risers.

Voltage: 300V but not printed on the jacket per UL 444 and UL 13.

Application: Communication Cable per NEC Art. 800 and UL 444; also for Class 3 or Class 2 Remote-Control, Signaling, and Power-Limited Circuits per NEC Art. 725 and UL 13.

Installation: Suitable for open cabling use in Risers as CMR, CL3R, or CL2R; also for open cabling in horizontal runs since rated higher than CM, CL3, CL2 respectively. Not for use in plenum spaces unless running in conduit.

Temp.: 0C to 60C minimum range, continuous operating ambient temperature.

QA: UL Listed as cable type CMR. Cable ratings shall be printed on the jacket. UL file number will be printed on the jacket alongside the gauge size and cable rating. UL label shall be displayed on every cable package near manufacturer's label.

Packaging: 1000ft or 500ft: spools, reels or boxes.

Appr. Manf: Manufacturer make and model shall be as listed in table 312UR.

Availability: Cable shall be available from local distribution known to project management.

Table 312UR Solid BC; PVC or PP Insul.; Multiple Pairs (N/S); Unshld O/A Riser Rated

Const	SDN	B	C	TAPPAN	BELDEN-WEST PENN	BELDEN
22/2pr	312UR.222	.185	17	R29909	251	-----
22/3pr	312UR.223	.197	23	R29908	252	-----
22/4pr	312UR.224	.213	28	R29907	253	-----
22/6pr	312UR.226	.247	40	R29400	-----	-----
22/9pr	312UR.229	.288	57	R29398	-----	-----
22/12pr	312UR.2212	.321	73	R29397	-----	-----
18/2pr	312UR.182	.244	33	R49915	-----	-----
18/3pr	312UR.183	.262	46	R49973	-----	-----

Manufacturer's makes vary. Diameters (Col. B) and weights (Col. C) are conservative and for engineering purposes only. Consult manufacturer's data for individual make & model information.

Column B: Nominal Cable Diameter in inches, for conduit fill calculations.
Column C: Nominal cable weight in pounds per 1000ft.

312UP: 300V Specification 312: Solid Bare Copper; Pairs (N/S) Unshielded O/A
Solid BC / LS-PVC Insul. / Multiple Pairs (N/S) / Unshielded (U) / Jacket. Plenum Rated (P).

Conductor: Solid Bare Copper.

Insulation: LS-PVC thickness according to UL 444.

Color Code: Pair (1) Black& Red, (2) Black&White, (3) Black&Green, (4) Black&Blue, (5) Black&Yellow, (6) Black&Brown, (7) Black&Orange, (8) Red&White, (9) Red&Green, (10) Red&Blue, (11) Red&Yellow, (12)Red&Brown.

Assembly: Insulated conductors twisted into pairs. Multiple pairs cabled together. Ripcord under the jacket to facilitate removal.

Shielding: None.

Jacket: LS-PVC. Thickness according to UL 444. Color: White or natural unless specified.

Diameter: Nominal diameter shall not exceed that in table 312UP Column B.

Flame test: Passes NFPA 262 (UL 910) for use in plenum spaces.

Voltage: 300V but not printed on the jacket per UL 444 and UL 13.

Application: Communication Cable per NEC Art. 800 and UL 444; also for Class 3 or Class 2 Remote-Control, Signaling, and Power-Limited Circuits per NEC Art. 725 and UL 13.

Installation: Suitable for open cabling use in plenum as CMP, CL3P, or CL2P; also for open cabling in risers or in horizontal runs since rated higher than those requirements..

Temp.: 0C to 60C minimum range, continuous operating ambient temperature.

QA: UL Listed as cable type CMP. Cable ratings shall be printed on the jacket. UL file number will be printed on the jacket alongside the gauge size and cable rating. UL label shall be displayed on every cable package near manufacturer's label.

Packaging: 1000ft or 500ft: spools, reels or boxes.

Appr. Manf: Manufacturer make and model shall be as listed in table 312UP.

Availability: Cable shall be available from local distribution known to project management.

Table 312UP Solid BC; LS-PVC Insul.; Multiple Pairs (N/S); Unshld O/A; Plenum Rated

Const	SDN	B	C	TAPPAN	BELDEN-WEST PENN	BELDEN
22/2pr	312UP.222	.188	19	P29914	------	6561UE
22/3pr	312UP.223	.193	22	P29647	------	-----
22/4pr	312UP.224	.201	27	P29646	------	-----
22/6pr	312UP.226	.235	39	P29612	------	-----
22/9pr	312UP.229	.276	56	P29632	------	-----
22/12pr	312UP.2212	.309	73	P29642	------	-----
18/2pr	312UP.182	.232	32	P49929	------	-----
18/3pr	312UP.183	.250	45	P49824	------	-----
18/4pr	312UP.184	.273	58	P49823	-----	-----

Manufacturer's makes vary. Diameters (Col. B) and weights (Col. C) are conservative and for engineering purposes only. Consult manufacturer's data for individual make & model information.

Column B: Nominal Cable Diameter in inches, for conduit fill calculations.
Column C: Nominal cable weight in pounds per 1000ft.

312SR: 300V Specification 312: Solid Bare Copper; Pairs (N/S) Shielded O/A
Solid BC / PVC or PP Insul. / Multiple Pairs (N/S) / Shielded Overall (S) / Jacket. Riser Rated (R).

Conductor: Solid Bare Copper

Insulation: PVC or Polypropylene, thickness according to UL 444.

Color Code: Pair (1) Black& Red, (2) Black&White, (3) Black&Green, (4) Black&Blue, (5) Black&Yellow, (6) Black&Brown, (7) Black&Orange, (8) Red&White, (9) Red&Green, (10) Red&Blue, (11) Red&Yellow, (12)Red&Brown.(Optional, First leg striped with the color of the second leg.)

Assembly: Insulated conductors twisted into pairs. Multiple pairs cabled together, and shielded overall. Ripcord under the jacket to facilitate removal.

Shielding: Aluminum Mylar Shielding Tape, Tinned Copper Drain Wire. Aluminum foil surface facing out; drain wire is outside the tape and in continuous contact with the foil.

Jacket: PVC. Thickness according to UL 444. Color: Gray unless specified.

Diameter: Nominal diameter shall not exceed that in table 312SR Column B.

Flame test: Passes UL 1666, for use in risers.

Voltage: 300V but not printed on the jacket per UL 444 and UL 13.

Application: Communication Cable per NEC Art. 800 and UL 444; also for Class 3 or Class 2 Remote-Control, Signaling, and Power-Limited Circuits per NEC Art. 725 and UL 13.

Installation: Suitable for open cabling use in Risers as CMR, CL3R, or CL2R; also for open cabling in horizontal runs since rated higher than CM, CL3, CL2 respectively.
Not for use in plenum spaces unless running in conduit.

Temp.: 0C to 60C minimum range, continuous operating ambient temperature.

QA: UL Listed as cable type CMR. Cable ratings shall be printed on the jacket.
UL file number will be printed on the jacket alongside the gauge size and cable rating.
UL label shall be displayed on every cable package near manufacturer's label.

Packaging: 1000ft or 500ft: spools, reels or boxes.
Appr. Manf: Manufacturer make and model shall be as listed in table 312SR.
Availability: Cable shall be available from local distribution known to project management.

Table 312SR Solid BC; PVC or PP Insul.; Multiple Pairs (N/S); Shielded O/A; Riser Rated

Const	SDN	B	C	TAPPAN	BELDEN-WEST PENN	BELDEN
22/2pr	312SR.222	.200	21	R29625	3251	5561FE
22/3pr	312SR.223	.212	26	R29516	3252	5562FE
22/4pr	312SR.224	.246	33	R29515	3253	5563FE
22/6pr	312SR.226	.247	42	R29514	-----	-----
22/9pr	312SR.229	.288	59	R29513	-----	-----
22/12pr	312SR.2212	.321	76	R29512	-----	-----
18/2pr	312SR.182	.244	36	R49916	-----	-----
18/3pr	312SR.183	.262	48	R49970	-----	-----

Manufacturer's makes vary. Diameters (Col. B) and weights (Col. C) are conservative and for engineering purposes only. Consult manufacturer's data for individual make & model information.

Column B: Nominal Cable Diameter in inches, for conduit fill calculations.
Column C: Nominal cable weight in pounds per 1000ft.

312SP: 300V Specification 312: Solid Bare Copper; Pairs (N/S) Shielded O/A
Solid BC / LS-PVC Insul. / Multiple Pairs (N/S) / Shielded Overall (S) / Jacket Plenum Rated (P).

Conductor: Solid Bare Copper.

Insulation: LS-PVC, thickness according to UL 444.

Color Code: Pair (1) Black& Red, (2) Black&White, (3) Black&Green, (4) Black&Blue, (5) Black&Yellow, (6) Black&Brown, (7) Black&Orange, (8) Red&White, (9) Red&Green, (10) Red&Blue, (11) Red&Yellow, (12)Red&Brown.

Assembly: Insulated conductors twisted into pairs. Multiple pairs cabled together, and shielded overall.. Ripcord under the jacket to facilitate removal.

Shielding: Aluminum Mylar Shielding Tape, Tinned Copper Drain Wire. Aluminum foil surface facing out; drain wire is outside the tape and in continuous contact with the foil.

Jacket: LS-PVC. Thickness according to UL 444. Color: White or natural unless specified.

Diameter: Nominal diameter shall not exceed that in table 312SP Column B.

Flame test: Passes NFPA 262 (UL 910) for use in plenum spaces.

Voltage: 300V but not printed on the jacket per UL 444 and UL 13.

Application: Communication Cable per NEC Art. 800 and UL 444. also for Class 3 or Class 2 Remote-Control, Signaling, and Power-Limited Circuits per NEC Art. 725 and UL 13..

Installation: Suitable for open cabling use in Plenums as CMP, CL3P, or CL2P; also for open cabling in risers and in horizontal runs since rated higher than those requirements..

Temp.: 0C to 60C minimum range, continuous operating ambient temperature

QA: UL Listed as cable type CMP. Cable ratings shall be printed on the jacket.
UL file number will be printed on the jacket alongside the gauge size and cable rating.
UL label shall be displayed on every cable package near manufacturer's label.

Packaging: 1000ft or 500ft: spools, reels or boxes.
Appr. Manf: Manufacturer make and model shall be as listed in table 312SP.
Availability: Cable shall be available from local distribution known to project management.

Table 312SP Solid BC; LS-PVC Insul.; Multiple Pairs (N/S); Shielded O/A; Plenum Rated

Const	SDN	B	C	TAPPAN	BELDEN
22/2pr	312SP.222	.193	23	P29796	6561FE
22/3pr	312SP.223	.197	23	P29641	-----
22/4pr	312SP.224	.201	30	P29640	-----
22/6pr	312SP.226	.235	40	P29638	-----
22/9pr	312SP.229	.276	59	P29635	-----
22/12pr	312SP.2212	.309	76	P29582	-----
18/2pr	312SP.182	.232	35	P49927	-----
18/3pr	312SP.183	.250	47	P49821	-----
18/4pr	312SP.184	.273	61	P49820	-----

Manufacturer's makes vary. Diameters (Col. B) and weights (Col. C) are conservative and for engineering purposes only. Consult manufacturer's data for individual make & model information.

Column B: Nominal Cable Diameter in inches, for conduit fill calculations.
Column C: Nominal cable weight in pounds per 1000ft.

314UR: 300V Specification 314: Solid Bare Copper; Pairs, EPS Unshielded O/A
Solid BC / PVC or PP Insul. / Multiple Pairs (EPS) / Unshielded O/A (U) / Jacket. Riser Rated (R).
Each Pair Shielded (EPS) – Isolated shields, one drain wire per pair.

Conductor: Solid Bare Copper.

Insulation: PVC or Polypropylene, thickness according to UL 444.

Color Code: Pair (1) Black& Red, (2) Black&White, (3) Black&Green, (4) Black&Blue, (5) Black&Yellow, (6) Black&Brown, (7) Black&Orange, (8) Red&White, (9) Red&Green. For coding of more pairs consult manufacturer. (Optional, First leg striped with the color of the second leg.)

Assembly: Insulated conductors twisted into pairs and individually shielded. Multiple pairs cabled together. Ripcord under the jacket to facilitate removal.

Shielding: Each pair: Aluminum-polyester shielding tape and a tinned copper drain wire. Aluminum foil surface facing in. Tape will be made with foil-free edges to ensure shield isolation or separate isolation tapes will be used. Overall: no overall shield.

Jacket: PVC. Thickness according to UL 444. Color: Gray unless specified.
Diameter: Nominal diameter shall not exceed that in table 314UR Column B.
Flame test: Passes UL 1666, for use in risers.
Voltage: 300V but not printed on the jacket per UL 444 and UL 13.
Application: Communication Cable per NEC Art. 800 and UL 444; also for Class 3 or Class 2 Remote-Control, Signaling, and Power-Limited Circuits per NEC Art. 725 and UL 13.

Installation: Suitable for open cabling use in Risers as CMR, CL3R, or CL2R; also for open cabling in horizontal runs since rated higher than CM, CL3, CL2 respectively.
Not for use in plenum spaces unless running in conduit.

Temp.: 0C to 60C minimum range, continuous operating ambient temperature.

QA: UL Listed as cable type CMR.
UL label shall be displayed on every cable package near manufacturer's label.

Packaging: 1000ft or 500ft: spools, reels or boxes.
Appr. Manf: Manufacturer make and model shall be as listed in table 314UR.
Availability: Cable shall be available from local distribution known to project management.

Table 314UR Solid BC; PVC or PP Insul.; EPS; Overall N/S; Riser Rated

Const	SDN	B	C	TAPPAN	BELDEN-WEST PENN
22/2pr	314UR.222	.195	21	R29453	420
22/3pr	314UR.223	.207	29	R29452	421
22/4pr	314UR.224	.234	37	R29476	429
22/6pr	314UR.226	.285	51	R29449	422
22/9pr	314UR.229	.336	75	R29446	423
22/10pr	314UR.2210	.358	83	R29445	-----
22/12pr	314UR.2212	.379	98	R29444	424
22/15pr	314UR.2215	.423	124	R29456	425
22/16pr	314UR.2216	.425	134	R29329	-----
18/4pr	314UR.184	.333	77	R49599	-----

Manufacturer's makes vary. Diameters (Col. B) and weights (Col. C) are conservative and for engineering purposes only. Consult manufacturer's data for individual make & model information.

Column B: Nominal Cable Diameter in inches, for conduit fill calculations.
Column C: Nominal cable weight in pounds per 1000ft

314UP: 300V Specification 314: Solid Bare Copper; Pairs, EPS Unshielded O/A
Solid BC / LS-PVC Insul. / Multiple Pairs (EPS) / Unshielded O/A (U) / Jacket Plenum Rated (P)
Each Pair Shielded (EPS) – Isolated shields, one drain wire per pair.

Conductor: Solid Bare Copper.

Insulation: LS-PVC, thickness according to UL 444.

Color Code: Pair (1) Black& Red, (2) Black&White, (3) Black&Green, (4) Black&Blue, (5) Black&Yellow, (6) Black&Brown, (7) Black&Orange, (8) Red&White, (9) Red&Green, . 10) Red&Blue, (11) Red&Yellow, (12)Red&Brown. For coding of more pairs consult manufacturer.

Assembly: Insulated conductors twisted into pairs with a drain wire and tape shield (foil side in.) Multiple pairs cabled together. Ripcord for jacket removal. PVC Jacket.

Shielding: Each pair: Aluminum-polyester shielding tape and a tinned copper drain wire. Aluminum foil surface facing in. Tape will be made with foil-free edges to ensure shield isolation or separate isolation tapes will be used. Overall: no overall shield.

Jacket: LS-PVC. Thickness according to UL 444. Color: Natural or white unless specified.

Diameter: Nominal diameter shall not exceed that in table 314UP Column B.

Flame test: Passes NFPA 262 (UL 910), for use in plenum spaces.

Voltage: 300V but not printed on the jacket per UL 444 and UL 13.

Application: Communication Cable per NEC Art. 800 and UL 444; also for Class 3 or Class 2 Remote-Control, Signaling, and Power-Limited Circuits per NEC Art. 725 and UL 13.

Installation: Suitable for open cabling use in Plenums as CMP, CL3P, or CL2P; also for open cabling in risers or in horizontal runs since rated higher than those requirements.

Temp.: 0C to 60C minimum range, continuous operating ambient temperature

QA: UL Listed as cable type CMP.
UL label shall be displayed on every cable package near manufacturer's label.

Packaging: 1000ft or 500ft: spools, reels or boxes.
Appr. Manf: Manufacturer make and model shall be as listed in table 314UP.
Availability: Cable shall be available from local distribution known to project management.

Table 314UP Solid BC; LS-PVC Insul.; EPS; Overall N/S; Plenum Rated

Const	SDN	B	C	TAPPAN
22/2pr	314UP.222	.186	21	P29656
22/3pr	314UP.223	.198	29	P29655
22/4pr	314UP.224	.228	38	P29654
22/6pr	314UP.226	.262	54	P29652
22/10pr	314UP.2210	.330	86	P29649
22/12pr	314UP.2212	.346	102	P29648
22/16pr	314UP.2216	.390	136	P29595
18/4pr	314UP.184	.318	71	P49806

Alternatives:
Specification 34;
Stranded Bare Copper
Table 34UP
or
Specification 324
Stranded Tinned Copper Upgrade Insulation
Table 324UP

Manufacturer's makes vary. Diameters (Col. B) and weights (Col. C) are conservative and for engineering purposes only. Consult manufacturer's data for individual make & model information.

Column B: Nominal Cable Diameter in inches, for conduit fill calculations.
Column C: Nominal cable weight in pounds per 1000ft.

315MR: 300V Specification 315: Solid Bare Copper; 1 pr + #/C **1 Pr Shielded (+) #/C**
Solid BC / PVC or PP Insul. / 1 Pr Shielded + Unshld. Conductors (M) / Jacket **Riser Rated (R).**
Mixed Shielded Cables – Combination 1 Shielded Pair + #/C.

Conductor: Solid bare copper, except as noted.

Insulation: PVC or Polypropylene, thickness according to UL 444.

Color Code: Black, Red (Inside the shield); White, Green outside the shield. Else, consult manufacturer.

Assembly: Black, Red insulated conductors twisted into a pair with a drain wire and tape shield (foil side in.) White and green (if present) insulated wires outside the shield. Ripcord for jacket removal. PVC Jacket.

Shielding: On the pair: Aluminum-polyester shielding tape and a tinned copper drain wire. Aluminum foil surface facing in. No overall shield

Jacket: PVC. Thickness according to UL 444. Color: Gray unless specified.

Diameter: Nominal diameter shall not exceed that in table 315MR Column B.

Flame test: Passes UL 1666, for use in risers.

Voltage: 300V but not printed on the jacket per UL 444 and UL 13.

Application: Communication Cable per NEC Art. 800 and UL 444; also for Class 3 or Class 2 Remote-Control, Signaling, and Power-Limited Circuits per NEC Art. 725 and UL 13.

Installation: Suitable for open cabling use in Risers as CMR, CL3R, or CL2R; also for open cabling in horizontal runs since rated higher than CM, CL3, CL2 respectively. Not for use in plenum spaces unless running in conduit.

Temp.: 0C to 60C minimum range, continuous operating ambient temperature.

QA: UL Listed as cable type CMR. Cable ratings shall be printed on the jacket. UL label shall be displayed on every cable package near manufacturer's label.

Packaging: 1000ft or 500ft: spools, reels or boxes.

Appr. Manf: Manufacturer make and model shall be as listed in table 315MR.

Availability: Cable shall be available from local distribution known to project management.

Table 315MR **Solid BC*; PVC or PP Insul.; Mixed Shielded + N/S; Riser Rated**

Const	SDN	B	C	TAPPAN	BELDEN-WEST PENN
1pr 22 SH + 1C 22	315MR.223	.130	15	R20726	351
1pr 22 SH + 2C 22	315MR.224	.140	17	R20727	355
1 pr 22 SH + 4C 22	315MR.226	.156	23	R21027	369
1 pr 22 SH + 2 pr 22 NS	315MR.226A	.198	25	R29519*	373
1pr 20 SH + 1C 20	315MR.203	.144	19	R30090	-----
1pr 20 SH + 2C 20	315MR.204	.156	23	R30091	-----
1pr 20 SH + 2 pr 18 NS	315MR.M46	.367	72	H91149	-----
1pr 18 SH + 1C 18	315MR.183	.167	26	R40875	-----
1pr 18 SH + 2C 18	315MR.184	.181	33	R40876	-----

*tinned copper

Manufacturer's makes vary. Diameters (Col. B) and weights (Col. C) are conservative and for engineering purposes only. Consult manufacturer's data for individual make & model information.

Column B: Nominal Cable Diameter in inches, for conduit fill calculations.
Column C: Nominal cable weight in pounds per 1000ft.

315MP: 300V Specification 315: Solid Bare Copper; 1 pr + #/C **1 Pr Shielded (+) #/C**
Solid BC / LS-PVC Insul. / 1 Pr Shielded + Unshld. Conductors (M) / Jacket. **Plenum Rated (P).**

Mixed Shielded Cables – Combination 1 Shielded Pair + #/C.

Conductor: Solid Bare Copper.

Insulation: LS-PVC, thickness according to UL 444.

Color Code: Black, Red (Inside the shield); White, Green outside the shield. Else, consult manufacturer.

Assembly: Black, Red insulated conductors twisted into a pair with a drain wire and tape shield (foil side in.) White and green (if present) insulated wires outside the shield. Ripcord for jacket removal. PVC Jacket.

Shielding: On the pair: Aluminum-polyester shielding tape and a tinned copper drain wire. Aluminum foil surface facing in. No overall shield

Jacket: LS-PVC. Thickness according to UL 444. Color: Natural or white unless specified.

Diameter: Nominal diameter shall not exceed that in table 315MP Column B.

Flame test: Passes NFPA 262 (UL 910), for use in plenum spaces.

Voltage: 300V but not printed on the jacket per UL 444 and UL 13.

Application: Communication Cable per NEC Art. 800 and UL 444 (300V); also for Class 3 or Class 2 Remote-Control, Signaling, and Power-Limited Circuits per NEC Art. 725 and UL 13.

Installation: Suitable for open cabling use in Plenums as CMP, CL3P, or CL2P; also for open cabling in Risers and in horizontal runs since rated higher than those requirements.

Temp.: 0C to 60C minimum range, continuous operating ambient temperature.

QA: UL Listed as cable type CMP. Cable ratings shall be printed on the jacket. UL file number will be printed on the jacket alongside the gauge size and cable rating. UL label shall be displayed on every cable package near manufacturer's label.

Packaging: 1000ft or 500ft: spools, reels or boxes.

Appr. Manf: Manufacturer make and model shall be as listed in table 315MP.

Availability: Cable shall be available from local distribution known to project management.

Table 315MP **Solid BC; LS-PVC Insul.; Mixed Shielded + N/S; Plenum Rated**

Const	SDN	B	C	TAPPAN	BELDEN-WEST PENN
1pr 22 SH + 1C 22	315MP.223	.143	15	P20740	-----
1pr 22 SH + 2C 22	315MP.224	.153	17	P20429	25355B
1pr 20 SH + 2C 20	315MP.204	.158	24	P30206	-----
1pr 18 SH + 2C 14 N/S	315MP.M54	.384	58	H91631	-----

Manufacturer's makes vary. Diameters (Col. B) and weights (Col. C) are conservative and for engineering purposes only. Consult manufacturer's data for individual make & model information

Column B: Nominal Cable Diameter in inches, for conduit fill calculations.
Column C: Nominal cable weight in pounds per 1000ft.

MASTER TABLE OF PUBLIC CABLE 1.0 PRODUCTS

PERIODIC TABLE OF PARTIAL CONTENTS – Detail for Stranded Tinned Copper, Upgrade Insulation

Cabled; Stranded Tinned Copper; Upgrade Insulation; **Specification 321**

Shielding	Installation	Spec / Table	Spec. Page	Table Page
Unshielded	Horizontal	321UH	82	83
	Plenum	321UP	84	85
Shielded	Horizontal	321SH	86	87
	Plenum	321SP	88	89 - 90

Multiple Unshielded Pairs; Stranded Tinned Copper; Upgrade Insulation; **Specification 322**

Shielding	Installation	Spec / Table	Spec. Page	Table Page
Unshld. Pairs / No OAS	Horizontal	322UH	92	93
	Plenum	322UP	94	95
Unshld. Pairs / OAS	Horizontal	322SH	96	97
	Plenum	322SP	98	99

Multiple Pairs; EPS/CD; Stranded Tinned Copper; Upgrade Insulation; **Specification 323**

Shielding	Installation	Table	Spec. Page	Table Page
EPS/CD	Horizontal	323UH	100	102
	Plenum	323UP	104	105

Multiple Pairs; EPS; Stranded Tinned Copper; Upgrade Insulation; **Specification 324**

Shielding	Installation	Table	Spec. Page	Table Page
EPS	Horizontal	324UH	101	102 – 103
	Plenum	324UP	106	107

RS-485; Str. TC, 120Ω Unshielded Pairs, Overall Foil + 90% Braid; **Specification 328A**

Shielding	Installation	Spec / Table	Spec. Page	Table Page
Pairs; OAS Foil + 95%TC	Horizontal	328AH	108	109
	Plenum	328AP	110	111

RS-232; Str. TC, 100Ω Unshielded Pairs, Overall Foil + 60% Braid; **Specification 328B**

Shielding	Installation	Spec / Table	Spec. Page	Table Page
Pairs; OAS Foil + 60%TC	Horizontal	328BH	112	113
	Plenum	328BP	114	115

EPS incl. RS-422; Str. TC, EPS Shielded Pairs, No Overall Shield; **Specification 328C**

Shielding	Installation	Spec / Table	Spec. Page	Table Page
EPS Pairs; No OAS	Horizontal	328CH	116	118
	Plenum	328CP	118	119

PERIODIC TABLE OF PARTIAL CONTENTS – Detail for Category Cable (partial)

Category 3 **Specification 333**

Shielding	Installation	Spec / Table	Spec. Page	Table Page
Category 3 UTP	Horizontal	333UH	122	122
	Plenum	333UP	123	123

Category 5e **Specification 335**

Shielding	Installation	Spec / Table	Spec. Page	Table Page
Category 5e UTP	Horizontal	335UH	124	124
	Plenum	335UP	125	125

321UH: 300V Specification 321: Stranded Tinned Copper; Cabled, Unshielded
Str. TC / Upgrade Insulation & Jacket / Cabled / Unshielded (U) / Jacketed. Horizontal Rated (H). some Riser Rated (R).

<u>Conductor</u>: Stranded tinned copper. Minimum number of strands per Table 321UH Column A. Higher stranding permitted, consult individual manufacturers' specifications.

<u>Insulation</u>: PVC. Thickness according to individual manufacturer's specification, being greater thickness than UL 444 or UL 13 minimum requirements.
For cables rated with an AWM style (with an asterisk in table 321UH) – minimum insulation thickness shall be according to manufacturer's UL AWM style sheets.

<u>Color Code</u>: Black, Red, White, Green. In two conductors cables, Black, White is permitted.

<u>Assembly</u>: Twisted together so that a 2 conductor cable is also considered 1 pair and 4 conductor cable is considered 1 quad. Ripcord under the jacket to facilitate removal.

<u>Shielding</u>: None.

<u>Jacket</u>: PVC. Thickness according to individual manufacturer's specification, being greater thickness than the UL 444 or UL 13 minimum requirements. . For cables rated with an AWM style (with an asterisk in table 321UH) – minimum jacket thickness shall be according to manufacturer's UL AWM style sheets.

<u>Diameter:</u> Nominal diameter s according to manufacturer's specifications. Nominal diameter shall not exceed that in table 321UH Column B.

<u>Flame test</u>: UL 1685 / NEC CMG or higher preferred. (*Items italicized in Table 321UH are VW-1 only.*)

<u>Voltage</u>: 300V (minimum) in-building rating. NEC CMG or higher. Where used as AWM, the voltage rating and use will be as for the associated AWM rating of the cable.

<u>Application</u>: 22 AWG through 16 AWG: Communication Cable per NEC Art. 800 and UL 444.
14 AWG – 12 AWG; Class 3 Remote-Control, Signaling, and Power-Limited Circuits per NEC Art. 725 and UL 13. Use as Appliance Wiring Material where and as listed.

<u>Installation</u>: As a minimum; suitable for open cabling use in horizontal cable runs in buildings, CMG listed or higher. Also for open cabling use in horizontal cable runs as CL2 or CL3 wiring since rated higher. Rating for riser use is optional and per individual manufacturers' cable specifications.

<u>Temp.</u>: 60C rated or higher. Where used as AWM, the temperature rating will be as for the associated AWM rating of the cable.

<u>QA</u>: UL Listed as cable type CMG or higher. (*Items italicized in Table 321UH are not CMG* .)

<u>Packaging</u>: 1000ft or 500ft: spools, reels or boxes.

<u>Appr. Manf</u>: Manufacturer and model shall be as listed in table 321UH, or approved equal.

<u>Availability</u>: Cable shall be available from local distribution known to project management.

Table 321UH Str. TC; Upgrade Insulation & Jacket; Cabled; Unshielded; Horizontal Rated or Riser Rated

Const	SDN	A	B	C	TAPPAN	GENERAL CABLE	BELDEN
22/2	321UH.222	7	.170	15	R20786*	C6348A	8442
22/4	321UH.224	7	.189	23	R20787*	C4063A*	8444*
20/2	321UH.202	7	.192	19	R30160*	C6351A	8205
20/4	321UH.204	7	.235	33	R30240*	C6353A*	9444*
18/2	321UH.182A	16	.210	25	R41006*	C6101A*	9740*
	321UH.182B	7	.234	28	R41058*	C2830A	8461
18/4	321UH.184	16	.258	48	R41184*	C2404A*	8489*
16/2	321UH.162	19	.274	43	R50120*	C2405A*	8471*
16/4	321UH.164	19	.376	88	R50094*	C2425A*	*Note 1*
14/2	321UH.142	19	.340	60	R60091*	C2409A*	8473*
14/4	321UH.144	19	.490	149	L60002*	C2430A*	*Note 2*
12/2	321UH.122	65	.390	85	R70058*	C2410A*	8477*

* Items with an asterisk also have an AWM rating.

Manufacturer's makes vary. Diameters (Col. B) and weights (Col. C) are conservative and for engineering purposes only. Consult manufacturer's data for individual make & model information

Column A: Minimum number of strands. Consult manufacturer's spec.'s for higher strands.
Column B: Cable Diameter in inches. Manufacturers' vary. Consult manufacturer's spec.'s.
Column C: Nominal cable weight in pounds per 1000ft

Note 1: 8620 is not CMG or otherwise rated (or listed) for in-building use. It is VW-1 rated.
Note 2: 8627 is not CMG or otherwise rated (or listed) for in-building use. It is VW-1 rated.

321UP: 300V Specification 321: Stranded Tinned Copper; Cabled **Unshielded**
Str. TC / Upgrade Insulation & Jacket / Cabled / Unshielded (U) / Jacket. **Plenum Rated (P).**

Conductor: Stranded tinned copper. Minimum number of strands per Table 321UP Column A. Higher stranding permitted, consult individual manufacturers' specifications.

Insulation: Material per Table 321UP (INS). Minimum insulation thickness according to UL 444.

Color Code: Black, Red, White, Green. Black, White for 2/C cables is permitted.

Assembly: Twisted together so that a 2 conductor cable is also considered 1 pair and 4 conductor cable is considered 1 quad.

Shielding: None.

Jacket: Material as per table 321UP (JKT.) Minimum thickness according to UL 444.

Diameter: Nominal diameter shall not exceed that in table 321UP Column B.

Flame test: Passes NFPA 262 (UL 910 Steiner Tunnel) flame test for use in plenum spaces; *except items in italics and Noted are VW-1 only.*

Voltage: 300V rating according to manufacturer's specification and UL 444.

Application: 22 AWG through 16 AWG: Communication Cable per NEC Art. 800 and UL 444.
14 AWG – 12 AWG; Class 3 or Class 2 Remote-Control, Signaling, and Power-Limited Circuits per NEC Art. 725 and UL 13.

Installation: Suitable for open cabling use in plenum spaces as CL2P, CL3P, or CMP.
Also for open cabling use in risers and in horizontal runs since rated higher than requirements.

Temp.: As per Table 321UP and manufacturer's specifications.

QA: UL Listed as cable type CL2P or higher such as CL3P or CMP (highest.)

Packaging: 1000ft or 500ft: spools, reels or boxes.

Appr. Manf: Manufacturer make and model shall be as listed in table 321UP.

Availability: Cable shall be available from local distribution known to project management.

Table 321UP **Str. TC; Upgrade Insul. & Jkt ; Cabled; N/S; Plenum Rated**

Const	SDN	A	INS	JKT	TEMP	B	C	TAPPAN	GENERAL CABLE	BELDEN
22/2	321UP.222A	7	1	3	60C	.114	10	M20042	-----	82442
	321UP.222B	7	1	1	200C	.148	9	T20027	-----	88442
22/4	321UP.224A	7	2	3	60C	.134	16	M20041	-----	82444
	321UP.224B	7	1	1	200C	.128	19	T20011	-----	88444
20/2	321UP.202A	7	2	2	150C	.150	14	K30008	-----	-----
	321UP.202B	7	2	2	150C	.185	23	-----	-----	*Note 1*
18/2	321UP.182A	19	1	3	75C	.154	17	M40096	C8122	82740
	321UP.182B	19	1	1	200C	.142	16	T40026	C8116	89740
18/4	321UP.184A	19	1	3	75C	.178	30	M40063	C8102	82489
	321UP.184B	19	1	1	200C	.170	30	T40037	-----	88489
16/2	321UP.162	19	1	1	200C	.162	25	T50055	-----	-----
	321UP.162	19	2	2	150C	.211	34	-----	-----	*Note 2*
16/4	321UP.164	19	1	1	200C	.191	46	T50012	-----	-----
14/2	321UP.142	19	1	1	200C	.188	36	T60008	-----	-----
14/4	321UP.144	19	1	1	200C	.223	67	T60009	-----	-----
12/2	321UP.122	19	1	1	200C	.232	54	T70012	-----	-----

Manufacturer's makes vary. Diameters (Col. B) and weights (Col. C) are conservative and for engineering purposes only. Consult manufacturer's data for individual make & model information.

Column A: Minimum number of strands. Consult manufacturer's spec.'s for higher strands.
INS: '1' indicates FEP Insulation. '2' indicates FEP or other fluoropolymer.
JKT: '1' indicates FEP. '2' indicates FEP or other fluoropolymer. '3' indicates LS-PVC
Column B: Cable Diameter in inches. Manufacturers' vary. Consult manufacturer's spec.'s
Column C: Nominal cable weight in pounds per 1000ft

Note 1: 85220 is not plenum or otherwise rated (or listed) for in-building use. It is VW-1 rated.
Note 2: 85221 is not plenum or otherwise rated (or listed) for in-building use. It is VW-1 rated.

321SH: 300V Specification 321: Stranded Tinned Copper; Cabled, Shielded
Str. TC / Upgrade Insulation & Jacket / Cabled / Shielded (S) / Jacket. Horizontal Rated (H). some Riser Rated (R).

Conductor: Stranded tinned copper. Minimum number of strands per Table 321SH Column A. Higher stranding permitted, consult individual manufacturers' specifications.

Insulation: Material per Table 321SH (INS). Minimum insulation thickness according to UL 444 or UL 13 as applicable; and according to manufacturer's listings and specifications.

Color Code: Black, Red, White, Green.
In two conductors cables, Black, Clear and Black, White are permitted.
Consult manufacturer's specifications.

Assembly: Twisted together so that a 2 conductor cable is also considered 1 pair and 4 conductor cable is considered 1 quad. Shielded.

Shielding: Aluminum Mylar Shielding Tape, Tinned Copper Drain Wire. Aluminum foil surface facing out; drain wire is outside the tape and in continuous contact with the foil. Except 321SH.222Bez and 321SH.182Bez will have the Aluminum foil surface facing in and the polyester face shall be bonded to the jacket. The drain wire for these two SDN's shall be inside the tape and in continuous contact with the aluminum face.

Jacket: PVC. Thickness according to individual manufacturer's specification. Minimum jacket thickness according to UL 444 or UL 13 as applicable. For cables rated with an AWM style (with an asterisk in table 321SH) – minimum jacket thickness shall be according to manufacturer's UL AWM style sheets.

Diameter: Nominal diameter s according to manufacturer's specifications. Nominal diameter shall not exceed that in table 321UH Column B.

Flame test: UL 1685 / NEC CMG or higher preferred.

Voltage: 300V except cables rated CL2 (only) are rated 150V. Where used as AWM, the voltage rating will be as for the associated AWM rating of the cable.

Application: 22 AWG through 16 AWG: Communication Cable per NEC Art. 800 and UL 444.
14 AWG – 12 AWG; Class 3 or Class 2 Remote-Control, Signaling, and Power-Limited Circuits per NEC Art. 725 and UL 13. For use as Appliance Wiring Material as applicable.

Installation: As a minimum; suitable for open cabling use in horizontal cable runs in buildings, CMG listed or higher. Also for open cabling use in horizontal cable runs as CL2 or CL3 wiring since rated higher. Rating for riser use is optional and per individual manufacturers' cable specifications.

Temp.: 60C rated r higher. Where used as AWM, the temperature rating will be as for the associated AWM rating of the cable.

QA: UL Listed as cable type CMG or higher. (*Items italicized in Table 321SH are not CMG.*)

Packaging: 1000ft or 500ft: spools, reels or boxes.

Appr. Manf: Manufacturer and model shall be as listed in table 321UH.

Availability: Cable shall be available from local distribution known to project management.

Table 321SH Str. TC; Upgrade Ins. & Jkt; Cabled; Shielded; Horizontal Rated or Riser Rated.
SDN numbers 221SH may be 150V maximum in-wall rated cables (CL2).
Consult manufacturer's specifications.

Const	SDN	A	INS	JKT	CAP	TEMP	B	C	TAPPAN	GENERAL CABLE	BELDEN West Penn	BELDEN
24/2	321SH.242A	7	2	1	2	60C	.168	14	L10051*	C2513A*	-----	8641*
	321SH.242B	7	1	1	1	60C	.146	12	L10006	-----	-----	-----
	321SH.242C	7	1	1	1	60C	.177	12	L10063	DS401	DA2401	1800B
22/2	321SH.222A	7	2	1	2	60C	.175	18	L20013*	C2514A*	77291	8761*
	321SH.222Bez	7	2	1	3	75C	.137	17	R21340	C2520A	(x)454	9451
	321SH.222C	7	1	1	1	75C	.174	17	L20003	-----	-----	-----
22/3	321SH.223	7	2	1	2	60C	.210	24	L20014*	C2528A*	-----	8771*
22/4	321SH.224A	19	2	1	2	60C	.245	37	L20050*	-----	-----	-----
	321SH.224B	19	2	1	2	75C	.257	43	-------	-----	-----	*Note 1.*
20/2	321SH.202A	7	2	1	2	60C	.190	22	L30009*	C2524A*	-----	-----
	321SH.202B	7	2	1	3	60C	.204	23	-----	-----	77292	8762*
20/3	321SH.203A	7	2	1	2	60C	.218	32	L30004*	C2525A*	-----	-----
	321SH.203B	7	2	1	3	60C	.218	32	-----	-----	-----	8772*
20/4	321SH.204	7	1	1	2	75C	.183	26	L30013	C2555A*	-----	-----
18/2	321SH.182A	16	2	1	2	60C	.222	29	L40008*	-----	77293	8760*
	321SH.182Bez	16	2	1	2	60C	.230	34	L40020*	C2521A*	-----	9460*
	321SH.182C	16	2	1	3	75C	.201	26	-----	C2534A*	-----	-----
18/3	321SH.183	16	2	1	2	60C	.246	39	L40009*	C2535A*	-----	8770*
18/4	321SH.184A	19	1	1	2	75C	.222	36	L40031	-----	-----	-----
	321SH.184B	19	3	1	3	75C	.246	46	A40132	-----	-----	-----
	321SH.184C	19	3	1	5	60C	.246	44	R40893*	C2543A*	-----	9418*
150V Maximum in-wall use if Rated CL2 (only) – (Items in small font). // ***Certain AWM ratings apply.**												
16/2	321SH.162	19	2	1	2	75C	.315	49	L50004*	C2536A*	77294	8719*
16/3	321SH.163	19	2	1	2	60C	.327	66	L50003*	C2537A*	-----	8618*
16/4	321SH.164A	19	1	1	2	75C	.256	55	L50009	-----	-----	-----
	321SH.164B	19	3	1	4	75C	.270	63	A50325	-----	-----	-----
14/2	221SH.142	19	2	1	2	60C	.355	71	L60001*	C2538A*	77295	8720*
12/2	221SH.122	19	2	1	3	80C	.412	102	L70001*	C2539A*	77296	8718*

* Items with an asterisk also have an AWM rating.

Manufacturer's makes vary. Diameters (Col. B) and weights (Col. C) are conservative and for engineering purposes only. Consult manufacturer's data for individual make & model information.

<u>'Bez' versions of 22/2 and 18/2</u>: The foil side of the shielding tape faces in; the drain wire is under the tape in contact with the aluminum side, and the polyester side is glued to the jacket for automated stripping bench equipment.

Column A: Minimum number of strands. Consult manufacturer's spec.'s for higher strands.
INS: '1'=Foam Polyethylene (FPE); '2'=Polyethylene or Polypropylene, '3'=PVC
JKT: '1'=PVC
CAP: Capacitance in pF/ft '1' = <16 '2'= 16-24 3"=25-35 '4'=36-49 '5'= 50 & up
Column B: Cable Diameter in inches. Manufacturers' vary. Consult manufacturer's spec.'s.
Column C: Nominal cable weight in pounds per 1000ft

Note 1: 8729 has Four Drain Wires of Copper-Coated Steel;
(3) being 23 AWG solid and (1) being 25 AWG stranded

321SP: 300V Specification 321: Stranded Tinned Copper; Cabled, Shielded
Str. TC / Upgrade Insulation & Jacket / Cabled / Shielded (S) / Jacket. Plenum Rated (P).

Conductor: Stranded Tinned Copper. Number of strands per Table 321SP Column A.

Insulation: Material per Table 321SP (INS). Minimum insulation thickness according to UL 444; and according to manufacturer's listings and specifications

Color Code: Black, Red, White, Green.
In two conductors cables, Black, White is permitted.
Consult manufacturer's specifications.

Assembly: Twisted together so that a 2 conductor cable is also considered 1 pair; 3 conductors is one triad; and 4 conductor cable is considered 1 quad. Shielded

Shielding: Aluminum-polyester Shielding Tape, Tinned Copper Drain Wire. Aluminum foil surface facing out; drain wire is outside the tape and in continuous contact with the foil.
Except 321SP.222Aez and 321SP.222Bez will have the Aluminum foil surface facing In. The drain wire for these two SDN's shall be inside the tape and in continuous contact with the aluminum face.

Jacket: Material as per table 321SP (JKT.) Minimum thickness according to UL 444.

Diameter: Nominal diameter shall not exceed that in table 321SP Column B.

Flame test: Passes NFPA 262 (UL 910 Steiner Tunnel) flame test for use in plenum spaces; *except items in italics and Noted are VW-1 only.*

Voltage: 300V rating according to manufacturer's specification and UL 444.

Application: Communication Cable per NEC Art. 800 and UL 444 or Class 3 or Class 2 Remote-Control, Signaling, and Power-Limited Circuits per NEC Art. 725 and UL 13.

Installation: Suitable for open cabling use in plenum spaces as CL2P, CL3P, or CMP.
Also for open cabling use in risers and in horizontal runs since rated higher than requirements.

Temp.: As per Table 321SP and manufacturer's specifications.

QA: UL Listed as cable type CL2P or higher such as CL3P or CMP (highest.)

Packaging: 1000ft or 500ft: spools, reels or boxes.

Appr. Manf: Manufacturer make and model shall be as listed in table 321SP.

Availability: Cable shall be available from local distribution known to project management.

Table 321SP appears on the next two pages.

Table 321SP to 20 AWG Str. TC; Upgrade Ins. & Jkt.; Cabled; Shielded; Plenum Rated.

Const	SDN	A	INS	JKT	CAP	TEMP	B	C	TAPPAN	GENERAL CABLE	BELDEN West Penn	BELDEN
24/2	321SP.242A	7	1	3	1	75C	.134	11	M10003	-----	-----	-----
	321SP.242B	7	1	3	1	75C	.170	13	M10009	DS401TS	DA252401	1801B
22/2	321SP.222Aez	7	1	3	2	75C	.135	14	M20100	-----	-----	-----
	321SP.222Bez	7	2	3	3	60C	.128	16	-----	C8124	D25454	9451P
	321SP.222C	7	2	3	3	75C	.120	13	M20027	C8126	D25291	82761
	321SP.222D	7	2	2	3	150C	.120	13	K20029	C8103	-----	87761
	321SP.222E	7	2	1	3	200C	.123	15	T20010	C8109	-----	88761
	321SP.222F	7	3	3	4	75C	.118	12	P20428	-----	-----	-----
22/3	321SP.223A	7	2	2	2	150C	.126	16	K20023	-----	-----	-----
	321SP.233B	7	3	3	4	75C	.126	15	P20399	-----	-----	-----
22/4	321SP.224A	7	2	2	2	150C	.139	19	K20024	-----	-----	-----
	321SP.224B	7	2	1	3	200C	.128	19	T20004	-----	-----	-----
	321SP.224C	7	3	3	4	75C	.137	19	P20448	-----	-----	-----
20/2	321SP.202A	7	2	3	3	60C	.126	15	M30011	-----	D25292	-----
	321SP.202B	7	2	2	3	150C	.130	15	K30001	-----	-----	-----
	321SP.202C	7	2	1	3	200C	.122	16	T30001	-----	-----	-----
	321SP.202D	7	2	1	3	200C	.160	25	T30012*	-----	-----	-----
	321SP.202E	7	2	2	3	150C	.182	29	-----	-----	-----	*Note 3*
	321SP.202F	7	3	3	4	75C	.130	14	P30088	-----	-----	-----
20/3	321SP.203A	7	2	2	3	150C	.139	20	K30002	-----	-----	-----
	321SP.203B	7	2	1	3	200C	.130	21	T30010	-----	-----	-----
	321SP.203C	7	2	2	3	150C	.193	30	-----	-----	-----	*Note 4*
	321SP.203D	7	3	3	4	75C	.139	19	P30089	-----	-----	-----
20/4	321SP.204A	7	2	2	3	150C	.153	25	K30003	-----	-----	-----
	321SP.204B	7	3	3	4	75C	.151	23	P30090	-----	-----	-----

(Continues on next page)

* Items with an asterisk also have an AWM rating.

'Aez' and 'Bez' versions of 22/2: The foil side of the shielding tape faces in; the drain wire is under the tape in contact with the aluminum side; the tape strips with the jacket; for alternate stripping methods.

Manufacturer's makes vary. Diameters (Col. B) and weights (Col. C) are conservative and for engineering purposes only. Consult manufacturer's data for individual make & model information.

Column A: Minimum number of strands. Consult manufacturer's spec.'s for higher strands.
INS: '1'=Foam FEP or Foam Fluoropolymer, '2'=FEP or other fluoropolymer, '3'=LS-PVC
JKT: '1'=FEP. '2'=other fluoropolymer. '3'= LS-PVC
CAP: Capacitance in pF/ft '1' = <16 '2'= 16-24 '3"=25-35 '4'=36-49 '5'= 50 & up
Column B: Cable Diameter in inches. Manufacturers' vary. Consult manufacturer's spec.'s.
Column C: Nominal cable weight in pounds per 1000ft

Note 3: 85230 is not CMP or otherwise rated (or listed) for in-building use. It is VW-1 rated.
Note 4: 85240 is not CMP or otherwise rated (or listed) for in-building use. It is VW-1 rated.

Table 321SP(Continued) **Str. TC; Upgrade Ins. & Jkt.; Cabled; Shielded; Plenum Rated.**

Const	SDN	A	INS	JKT	CAP	TEMP	B	C	TAPPAN	GENERAL CABLE	BELDEN West Penn	BELDEN
18/2	321SP.182A	19	1	3	1	75C	.222	27	M40097	-----	-----	-----
	321SP.182B	19	2	3	2	75C	.178	24	M40033	-----	-----	-----
	321SP.182C	7	2	1	3	200C	.180	32	T40064*	-----	-----	-----
	321SP.182D	19	2	1	4	200C	.144	21	T40007	-----	-----	-----
	321SP.182E	19	2	3	5	75C	.154	22	P41111	C8123	D25293	82760
	321SP.182F	19	2	2	5	150C	.154	22	K40010	C8104	-----	87760
	321SP.182G	19	2	1	5	200C	.154	24	T40030	C8101	-----	88760
	321SP.182H	19	3	3	5	75C	.154	19	P40668	-----	-----	-----
18/3	321SP.183A	7	2	2	3	150C	.178	29	K40005	-----	-----	-----
	321SP.183B	7	2	1	3	200C	.191	40	T40065*	-----	-----	-----
	321SP.183C	19	2	1	5	200C	.156	29	T40047	C8106	-----	88770
	321SP.183D	19	2	1	5	200C	.152	28	T40008	-----	-----	-----
	321SP.183E	19	3	3	5	75C	.165	26	P41211	-----	-----	-----
18/4	321SP.184A	19	2	2	3	150C	.177	34	K40008	-----	-----	-----
	321SP.184	19	2	3	3	75C	.172	37	P41595	-----	-----	82418
	321SP.184	19	2	1	3	200C	.180	37	T40036	C8114	-----	89418
	321SP.184	19	2	1	4	200C	.167	34	T40031	-----	-----	-----
16/2	321SP.162A	19	2	2	3	150C	.170	28	K50003	-----	D25294	-----
	321SP.162B	19	2	1	3	200C	.208	46	T50056*	-----	-----	-----
	321SP.162C	19	2	1	4	200C	.162	32	T50036	-----	-----	-----
	321SP.162D	19	2	2	4	150C	.210	38	-----	-----	-----	*Note 1*
	321SP.162E	19	3	3	5	75C	.178	28	P50159	-----	-----	-----
16/3	321SP.163A	19	2	2	3	150C	.183	38	K50004	-----	-----	-----
	321SP.163B	19	2	1	3	200C	.221	58	T50057*	-----	-----	-----
	321SP.163C	19	2	2	3	150C	.223	49	-----	-----	-----	*Note 2*
	321SP.163D	19	2	1	4	200C	.174	42	T50034	-----	-----	-----
16/4	321SP.164A	19	2	2	3	150C	.202	49	K50005	-----	-----	-----
	321SP.164B	19	2	1	3	200C	.232	68	T50058*	-----	-----	-----
	321SP.164C	19	2	1	4	200C	.191	53	T50001	-----	-----	-----

* Items with an asterisk also have an AWM rating.

Manufacturer's makes vary. Diameters (Col. B) and weights (Col. C) are conservative and for engineering purposes only. Consult manufacturer's data for individual make & model information.

Column A:	Minimum number of strands. Consult manufacturer's spec.'s for higher strands.
INS:	'1'=Foam FEP or Foam Fluoropolymer, '2'=FEP or other fluoropolymer, '3'=LS-PVC
JKT:	'1'=FEP. '2'=other fluoropolymer. '3'= LS-PVC
CAP:	Capacitance in pF/ft '1' = <16 '2'= 16-24 '3"=25-35 '4'=36-49 '5'= 50 & up
Column B:	Cable Diameter in inches. Manufacturers' vary. Consult manufacturer's spec.'s.
Column C:	Nominal cable weight in pounds per 1000ft

Note 1: 85231 is not CMP or otherwise rated (or listed) for in-building use. It is VW-1 rated.
Note 2: 85241 is not CMP or otherwise rated (or listed) for in-building use. It is VW-1 rated.

PERIODIC TABLE OF PARTIAL CONTENTS – Detail for Stranded Tinned Copper, Upgrade Insulation

Stranded Tinned Copper; Upgrade Insulation; Cabled — Specification 321

Shielding	Installation	Spec / Table	Spec. Page	Table Page
Unshielded	Horizontal	321UH	82	83
	Plenum	321UP	84	85
Shielded	Horizontal	321SH	86	87
	Plenum	321SP	88	89 - 90

Stranded Tinned Copper; Upgrade Insulation; Multiple Unshielded Pairs — Specification 322

Shielding	Installation	Spec / Table	Spec. Page	Table Page
Unshld. Pairs / No OAS	Horizontal	322UH	92	93
	Plenum	322UP	94	95
Unshld. Pairs / OAS	Horizontal	322SH	96	97
	Plenum	322SP	98	99

Stranded Tinned Copper; Upgrade Insulation; Multiple Pairs; EPS/CD — Specification 323

Shielding	Installation	Table	Spec. Page	Table Page
EPS/CD	Horizontal	323UH	100	102
	Plenum	323UP	104	106

Stranded Tinned Copper; Upgrade Insulation; Multiple Pairs; EPS — Specification 324

Shielding	Installation	Table	Spec. Page	Table Page
EPS	Horizontal	324UH	101	102 - 103
	Plenum	324UP	105	107

RS-485; Str. TC, 120Ω Unshielded Pairs, Overall Foil + 90% Braid; — Specification 328A

Shielding	Installation	Spec / Table	Spec. Page	Table Page
Pairs; OAS Foil + 95%TC	Horizontal	328AH	108	109
	Plenum	328AP	110	111

RS-232; Str. TC, 100Ω Unshielded Pairs, Overall Foil + 60% Braid; — Specification 328B

Shielding	Installation	Spec / Table	Spec. Page	Table Page
Pairs; OAS Foil + 60%TC	Horizontal	328BH	112	113
	Plenum	328BP	114	115

EPS incl. RS-422; Str. TC, EPS Shielded Pairs, No Overall Shield; — Specification 328C

Shielding	Installation	Spec / Table	Spec. Page	Table Page
EPS Pairs; No OAS	Horizontal	328CH	116	118
	Plenum	328CP	118	119

PERIODIC TABLE OF PARTIAL CONTENTS – Detail for Category Cable (partial)

Category 3 — Specification 333

Shielding	Installation	Spec / Table	Spec. Page	Table Page
Category 3 UTP	Horizontal	333UH	122	122
	Plenum	333UP	123	123

Category 5e — Specification 335

Shielding	Installation	Spec / Table	Spec. Page	Table Page
Category 5e UTP	Horizontal	335UH	124	124
	Plenum	335UP	125	125

322UH: 300V Specification 322: Stranded Tinned Copper; Pairs (N/S) Unshielded
Str. TC / Upgrade Insulation & Jacket / Multiple Pairs / Unshielded (U) / Jacket. Horizontal Rated (H). or Riser Rated (R).

Conductor: Stranded Tinned Copper.

Insulation: PVC, thickness according to approved manufacturer's UL listing and specification. Thickness shall be according to UL 444 as a minimum, and may be thicker where an AWM type is indicated. (Asterisked items in Table 322UH.)

Color Code: According to approved manufacturer's specification; preferably and typically:
Pair (1) Black& Red, (2) Black&White, (3) Black&Green, (4) Black&Blue, (5) Black&Yellow, (6) Black&Brown, (7) Black&Orange, (8) Red&White, (9) Red&Green; per manufacturer's specification for more pairs.

Or: for 2 pairs: (1) Black&Red; (2)White&Green

Assembly: Insulated conductors twisted into pairs. Multiple pairs cabled together. Ripcord under the jacket to facilitate removal.

Shielding: None.

Jacket: PVC. Thickness according to UL 444 as a minimm. and may be thicker where an AWM type is indicated. (Asterisked items in Table 322UH.) . Color: Gray unless specified.

Diameter: Nominal diameter shall not exceed that in table 322UH Column B.

Flame test: UL 1685.

Voltage: 300V but not printed on the jacket per UL 444. -- AWM Rating 150V. 150V Maximum voltage in AWM applications.

Application: Communication Cable per NEC Art. 800 and UL 444.; also for Class 3 or Class 2 Remote-Control, Signaling, and Power-Limited Circuits per NEC Art. 725. For use as Appliance Wiring Material when printed with an AWM style number; and if so is then limited to the conditions defined for that AWM style.

Installation: Suitable for open cabling use in horizontal runs as cable type CM, CL3, or CL2 respectively. Not for use in risers unless listed as cable type CMR, CL3R, or CL2R. Not for use in plenum spaces unless running in conduit.

Temp.: 0C to 60C or as according to table 322UH, for NEC installations.

QA: UL Listed as cable type CM. Cable ratings shall be printed on the jacket. UL file number will be printed on the jacket alongside the gauge size and cable rating. UL label shall be displayed on every cable package near manufacturer's label.

Packaging: 1000ft or 500ft: spools, reels or boxes.

Appr. Manf: Manufacturer make and model shall be as listed in table 322UH.

Availability: Cable shall be available from local distribution known to project management.

Table 322UH Str. TC; Upgrade Insulation; Pairs (N/S); Unshielded OA; Horizontal Rated. or Riser Rated

Const	SDN	INS	JKT	CAP	TEMP	B	C	TAPPAN	GENERAL CABLE	BELDEN
22/2pr	322UH.222A	3	1	3	80C	.233	27	R29459*	C6010A*	9744*
	322UH.222B	3	1	3	75C	.212	20	R29330	-----	
22/3pr	322UH.223A	3	1	3	80C	.250	36	R29561*	C6014A*	9745*
	322UH.223B	3	1	3	75C	.216	25	R29798	-----	-----
22/4pr	322UH.224A	3	1	3	80C	.281	42	R29464*	C6015A*	9746*
	322UH.224B	3	1	3	75C	.229	32	R29797	-----	-----
22/6pr	322UH.226A	3	1	3	80C	.320	59	R29559*	C6017A*	8747*
	322UH.226B	3	1	3	75C	.265	45	R29795	-----	-----
22/9pr	322UH.229A	3	1	3	80C	.389	84	R29558*	C6019A*	8748*
	322UH.229B	3	1	3	75C	.320	67	R29792	-----	-----
22/12pr	322UH.2212A	3	1	3	80C	.425	109	R29557*	C6032A*	9747*
	322UH.2212B	3	1	3	75C	.356	86	R29789	-----	-----
22/15pr	322UH.2215A	3	1	3	80C	.443	125	R29556*	C6026A*	8749*
	322UH.2215B	3	1	3	75C	.393	109	R29788	-----	-----
22/19pr	322UH.2219	3	1	3	80C	.505	159	R29555*	-----	9748*
22/27pr	322UH.2227	3	1	3	80C	.575	221	R29553*	-----	8750*

* Items with an asterisk have an AWM rating, some being associated with 150V for AWM use.

Manufacturer's makes vary. Diameters (Col. B) and weights (Col. C) are conservative and for engineering purposes only. Consult manufacturer's data for individual make & model information.

INS: '1'=Foam Polyethylene (FPE); '2'=Polyethylene or Polypropylene, '3'=PVC
JKT: '1'=PVC
CAP: Capacitance in pF/ft '1' = <16 '2'= 16-24 3"=25-35 '4'=36-49 '5'= 50 & up
Column B: Cable Diameter in inches. Manufacturers' vary. Consult manufacturer's spec.'s.
Column C: Nominal cable weight in pounds per 1000ft

322UP: 300V Specification 322: Stranded Tinned Copper; Pairs (N/S) Unshielded
Str. TC / Upgrade Insulation & Jacket / Multiple Pairs / Unshielded (U) / Jacketed. Plenum Rated (P).

Conductor: Stranded Tinned Bare Copper.

Insulation: Material as indicated in Table 322UP. Thickness shall be according to UL 444 as a minimum. Insulation thickness according to approved manufacturer's specification.

Color Code: Pair (1) Black& Red, (2) Black&White, (3) Black&Green, (4) Black&Blue, (5) Black&Yellow, (6) Black&Brown, (7) Black&Orange, (8) Red&White, (9) Red&Green; per manufacturer's specification for more pairs.

Assembly: Insulated conductors twisted into pairs. Multiple pairs cabled together. Ripcord under the jacket to facilitate removal.

Shielding: None.

Jacket: Material as indicated in Table 322UP Thickness according to UL 444. Color: White unless specified.

Diameter: Nominal diameter shall not exceed that in table 322UP Column B.

Flame test: Passes NFPA 262 (previously UL 910)

Voltage: 300V but not printed on the jacket per UL 444.

Application: Communication Cable per NEC Art. 800 and UL 444. also for Class 3 or Class 2 Remote-Control, Signaling, and Power-Limited Circuits per NEC Art. 725.

Installation: Suitable for open cabling use in plenum air-handling spaces. Also for open cabling in risers and in_horizontal runs since rated higher than those requirements.

Temp.: 0C to 60C or as shown in table 322UP.

QA: UL Listed as cable type CMP. Cable ratings shall be printed on the jacket. UL file number will be printed on the jacket alongside the gauge size and cable rating. UL label shall be displayed on every cable package near manufacturer's label.

Packaging: 1000ft or 500ft: spools, reels or boxes.

Appr. Manf: Manufacturer make and model shall be as listed in table 322UP.

Availability: Cable shall be available from local distribution known to project management.

Table 322UP Str. TC; Upgrade Insulation; Pairs (N/S) ; Unshielded OA; Plenum Rated.

Const	SDN	INS	JKT	CAP	TEMP	B	C	TAPPAN	BELDEN
22/2pr	322UP.222A	2	1	2	200C	.169	16	-----	88741
	322UP.222B	2	3	2	60C	.179	17	-----	82741
	322UP.222C	3	3	2	75C	.186	18	P29745	-----
22/3pr	322UP.223B	2	3	2	60C	.191	26	P29577	82742
	322UP.223C	3	3	2	75C	.199	24	P29744	-----
22/4pr	322UP.224A	2	1	2	200C	.200	29	-----	88757
	322UP.224B	2	3	2	60C	.210	28	-----	82757
	322UP.224C	3	3	2	75C	.217	31	P29743	-----
22/6pr	322UP.226B	2	3	2	60C	.238	43	P29575	82743
	322UP.226C	3	3	2	75C	.253	44	P29741	-----
22/12pr	322UP.2212	3	3	2	75C	.342	86	P29738	-----
22/15pr	322UP.2215	3	3	2	75C	.369	105	P29737	-----

Manufacturer's makes vary. Diameters (Col. B) and weights (Col. C) are conservative and for engineering purposes only. Consult manufacturer's data for individual make & model information.

INS: '1'=Foam FEP or Foam Fluoropolymer, '2'=FEP or other fluoropolymer, '3'=LS-PVC
JKT: '1'=FEP. '2'=other fluoropolymer. '3'= LS-PVC
CAP: Capacitance in pF/ft '1' = <16 '2'= 16-24 '3"=25-35 '4'=36-49 '5'= 50 & up
Column B: Cable Diameter in inches. Manufacturers' vary. Consult manufacturer's spec.'s.
Column C: Nominal cable weight in pounds per 1000ft

322SH: 300V Specification 322: Stranded Tinned Copper; Pairs (N/S)
Str TC / Upgrade Insulation & Jacket / Multiple Pairs / Shielded OA (S) / Jacket

Shielded Overall
Horizontal Rated (H).
or Riser Rated (R).

Conductor:	Stranded Tinned Copper.
Insulation:	Material according to table 322SH. Thickness according to manufacturer's UL listing and specification. Thickness shall be according to UL 444 as a minimum, and may be thicker where an AWM type is indicated. (Asterisked items in Table 322SH.)
Color Code:	According to approved manufacturer's specification; preferably and typically: Pair (1) Black& Red, (2) Black&White, (3) Black&Green, (4) Black&Blue, (5) Black&Yellow, (6) Black&Brown, (7) Black&Orange, (8) Red&White, (9) Red&Green; per manufacturer's specification for more pairs.
Assembly:	Insulated conductors twisted into pairs. Multiple pairs cabled together. Overall shield. Ripcord under the jacket to facilitate removal.
Shielding:	Aluminum-polyester Shielding Tape, Tinned Copper Drain Wire. Aluminum foil surface facing out; drain wire is outside the tape and in continuous contact with the foil.
Jacket:	PVC. Thickness according to UL 444 as a minimm. and may be thicker where an AWM type is indicated. (Asterisked items in Table 322SH.) . Color: Gray unless specified.
Diameter:	Nominal diameter shall not exceed that in table 322SH Column B.
Flame test:	UL 1685.
Voltage:	300V but not printed on the jacket per UL 444; for NEC concealed wiring installation. AWM applications; voltage rating according to AWM style and jacket print.
Application:	Communication Cable per NEC Art. 800 and UL 444.; also for Class 3 or Class 2 Remote-Control, Signaling, and Power-Limited Circuits per NEC Art. 725. For use as Appliance Wiring Material when printed with an AWM style number; and if so is then limited to the conditions defined for that AWM style.
Installation:	Suitable for open cabling use in horizontal runs as cable type CM, CL3, or CL2 respectively. Not for use in risers unless listed as cable type CMR, CL3R, or CL2R. Not for use in plenum spaces unless running in conduit.
Temp.:	0C to 60C or as shown in table 322SH.
QA:	UL Listed as cable type CM. Cable ratings shall be printed on the jacket. UL file number will be printed on the jacket alongside the gauge size and cable rating. UL label shall be displayed on every cable package near manufacturer's label.
Packaging:	1000ft or 500ft: spools, reels or boxes.
Appr. Manf:	Manufacturer make and model shall be as listed in table 322SH.
Availability:	Cable shall be available from local distribution known to project management.

Table 322SH Str. TC; Upgrade Insulation; Pairs N/S; Shielded O/A; Horizontal Rated.

Const	SDN	INS	JKT	CAP	TEMP	B	C	TAPPAN	BELDEN West Penn
22/2pr	322SH.222A	1	1	2	75C	.232	26	L29989	-----
	322SH.222B	2	1	3	60C	.235	27	-----	D3651
	322SH.222C	3	1	3	80C	.236	29	R29609*	-----
	322SH.222D	3	1	4	75C	.198	21	R29782	-----
22/3pr	322SH.223A	1	1	2	75C	.246	33	L29988	-----
	322SH.223B	2	1	3	60C	.248	34	-----	D3652
	322SH.223C	3	1	3	80C	.250	36	R29608*	-----
	322SH.223D	3	1	4	75C	.211	28	R29781	-----
22/4pr	322SH.224A	1	1	2	75C	.266	40	L29987	-----
	322SH.224B	2	1	3	60C	.268	41	-----	D3653
	322SH.224C	3	1	3	80C	.270	44	R29607*	-----
	322SH.224D	3	1	4	75C	.229	34	R29780	-----
22/6pr	322SH.226A	1	1	2	75C	.305	54	L29986	----
	322SH.226B	2	1	3	60C	.307	55	-----	D3654
	322SH.226C	3	1	3	80C	.309	59	R29425*	-----
	322SH.226D	3	1	4	75C	.265	47	R29778	-----
22/9pr	322SH.229C	3	1	3	80C	.360	82	R29424*	-----
	322SH.229D	3	1	4	75C	.320	70	R29775	-----
22/12pr	322SH.2212D	3	1	4	75C	.356	89	R29772	-----
22/15pr	322SH.2215C	3	1	3	80C	.445	127	R29423*	-----
	322SH.2215D	3	1	4	75C	.393	112	R29771	-----
22/19pr	322SH.2219	3	1	3	80C	.438	127	R29422*	-----

* Items with an asterisk have an AWM rating, some being associated with 150V for AWM use.

Manufacturer's makes vary. Diameters (Col. B) and weights (Col. C) are conservative and for engineering purposes only. Consult manufacturer's data for individual make & model information.

INS: '1'=Foam Polyethylene (FPE); '2'=Polyethylene or Polypropylene, '3'=PVC
JKT: '1'=PVC
CAP: Capacitance in pF/ft '1' = <16 '2'= 16-24 3"=25-35 '4'=36-49 '5'= 50 & up
Column B: Cable Diameter in inches. Manufacturers' vary. Consult manufacturer's spec.'s.
Column C: Nominal cable weight in pounds per 1000ft

322SP: 300V Specification 322: Stranded Tinned Copper; Pairs (N/S) Shielded Overall
Str TC / Upgrade Insulation & Jacket / Multiple Pairs / Shielded OA (S) / Jkt. Plenum Rated (P).

Conductor:	Stranded Tinned Bare Copper.
Insulation:	Material as indicated in Table 322SP. Thickness shall be according to UL 444 as a minimum. Insulation thickness according to approved manufacturer's specification.
Color Code:	Pair (1) Black& Red, (2) Black&White, (3) Black&Green, (4) Black&Blue, (5) Black&Yellow, (6) Black&Brown, (7) Black&Orange, (8) Red&White, (9) Red&Green; per manufacturer's specification for more pairs.
Assembly:	Insulated conductors twisted into pairs. Multiple pairs cabled together. Overall shield. Ripcord under the jacket to facilitate removal.
Shielding:	Aluminum-polyester Shielding Tape, Tinned Copper Drain Wire. Aluminum foil surface facing out; drain wire is outside the tape and in continuous contact with the foil.
Jacket:	Material as indicated in Table 322SP Thickness according to UL 444 (minimum). Color: White unless specified.
Diameter:	Nominal diameter shall not exceed that in table 322SP Column B.
Flame test:	Passes NFPA 262 (previously UL 910)
Voltage:	300V but not printed on the jacket per UL 444.
Application:	Communication Cable per NEC Art. 800 and UL 444. also for Class 3 or Class 2 Remote-Control, Signaling, and Power-Limited Circuits per NEC Art. 725.
Installation:	Suitable for open cabling use in plenum air-handling spaces. Also for open cabling in risers and in_horizontal runs since rated higher than those requirements.
Temp.:	0C to 60C or as shown in table 322SP.
QA:	UL Listed as cable type CMP. Cable ratings shall be printed on the jacket. UL file number will be printed on the jacket alongside the gauge size and cable rating. UL label shall be displayed on every cable package near manufacturer's label.
Packaging:	1000ft or 500ft: spools, reels or boxes.
Appr. Manf:	Manufacturer make and model shall be as listed in table 322SP.
Availability:	Cable shall be available from local distribution known to project management.

Table 322SP Str. TC; Up Grade Insulation; Pairs, (N/S); Shielded O/A; Plenum Rated.

Const	SDN	INS	JKT	CAP	TEMP	B	C	TAPPAN	GENERAL CABLE	BELDEN West Penn
22/2pr	322SP.222B	2	1	2	200C	.222	34	T29990*	-----	-----
	322SP.222C	2	2	2	150C	.192	21	K29984	C3205	----
	322SP.222D	2	3	3	60C	.191	22	-----	-----	D253651
	322SP.222E	3	3	4	75C	.186	20	P29733	-----	-----
22/3pr	322SP.223B	2	1	2	200C	.236	42	T29989*	-----	-----
	322SP.223C	2	2	2	150C	.206	29	K29983	C3206	-----
	322SP.223D	2	3	3	60C	.204	30	-----	-----	D253652
	322SP.223E	3	3	4	75C	.199	27	P29732	-----	
22/4pr	322SP.224A	1	3	1	75C	.276	40	M29957	-----	-----
	322SP.224B	2	1	2	200C	.256	50	T29988*	-----	-----
	322SP.224C	2	2	2	150C	.226	36	K29981	C3207	-----
	322SP.224D	2	3	2	60C	.224	37	-----	-----	D253653
	322SP.224E	3	3	4	75C	.217	33	P29731	-----	-----
22/6pr	322SP.226B	2	1	2	200C	.275	59	T29987*	-----	-----
	322SP.226C	2	2	2	150C	.237	46	-----	C3208	-----
22/8pr	322SP.228B	2	1	2	200C	.324	84	T29984*	-----	-----
	322SP.228E	3	3	4	75C	.284	61	P29729	-----	-----
22/12pr	322SP.2212	3	3	4	75C	.342	89	P29727	-----	-----

* Items with an asterisk have an AWM rating.

Manufacturer's makes vary. Diameters (Col. B) and weights (Col. C) are conservative and for engineering purposes only. Consult manufacturer's data for individual make & model information.

INS: '1'=Foam FEP or Foam Fluoropolymer, '2'=FEP or other fluoropolymer, '3'=LS-PVC
JKT: '1'=FEP. '2'=other fluoropolymer. '3'= LS-PVC
CAP: Capacitance in pF/ft '1' = <16 '2'= 16-24 '3''=25-35 '4'=36-49 '5'= 50 & up
Column B: Cable Diameter in inches. Manufacturers' vary. Consult manufacturer's spec.'s.
Column C: Nominal cable weight in pounds per 1000ft

323UH: 300V Specification 323: Stranded Tinned Copper; Pairs EPS/CD **Unshielded O/A**
Str TC / Upgr. Insul & Jkt / Multiple Pairs (EPS/CD) / Unshld O/A (U) / Jacket **Horizontal Rated (H). or Riser Rated (R).**

Each Pair Shielded (EPS) / Common Drain Wire (CD)

Conductor:	Stranded Tinned Copper
Insulation:	Material according to Table 323UH. Thickness according to UL 444.
Color Code:	Pair (1) Black& Red, (2) White & Green
Assembly:	Insulated conductors twisted into pairs and shielded (foil side out.). Multiple pairs cabled together with a common drain wire. Ripcord for jacket removal. PVC Jacket.
Shielding:	Each pair: Aluminum-polyester shielding tape. Aluminum foil surface facing out. Tinned Copper Drain Wire. Drain wire is outside the tapes and in continuous contact with the both foil shields.
Jacket:	PVC. Thickness according to UL 444. Color: Gray unless specified.
Diameter:	Nominal diameter shall not exceed that in table 323UH Column B.
Flame test:	UL 1685.
Voltage:	300V but not printed on the jacket per UL 444..
Application:	Communication Cable per NEC Art. 800 and UL 444; also for Class 3 or Class 2 Remote-Control, Signaling, and Power-Limited Circuits per NEC Art. 725.
Installation:	Suitable for open cabling use in horizontal concealed spaces per the NEC as CM, CL3, or CL2.. Not for use in risers or in plenum spaces unless running in conduit.
Temp.:	0C to 60C, or as indicated in Table 323UH. Consult manufacturer's specifications.
QA:	UL Listed as cable type CM. Cable ratings shall be printed on the jacket. UL file number will be printed on the jacket alongside the gauge size and cable rating. UL label shall be displayed on every cable package near manufacturer's label.
Packaging:	1000ft or 500ft: spools, reels or boxes.
Appr. Manf:	Manufacturer make and model shall be as listed in table 323UH.
Availability:	Cable shall be available from local distribution known to project management.

Table 323UH appears on page

13410-102

324UH: 300V Specification 324: Stranded Tinned Copper; Shielded Pairs
Str TC / Upgrade Insulation & Jacket / Shielded Pairs / no OA Shld (U) / Jacket.

EPS
Horizontal Rated (H).
or Riser Rated (R).

Each Pair Shielded (EPS) / Individual Drain Wires (1) per Pair

<u>Conductor</u>: Stranded Tinned Copper. – 7 strands..

<u>Insulation</u>: PVC or Polypropylene, thickness according to UL 444.

<u>Color Code</u>: Pairs shall be distinctly differentiated from each other. Color code per manufactures' specification. Preferred: Pair (1) Black& Red, (2) Black&White, (3) Black&Green, (4) Black&Blue, (5) Black&Yellow, (6) Black&Brown, (7) Black&Orange, (8) Red&White, (9) Red&Green. Consult manufactures' specifcications

<u>Assembly</u>: Insulated conductors twisted into pairs with a drain wire and tape shield (foil side in.) Multiple pairs cabled together. Shields are electrically isolated from each other.
Ripcord for jacket removal. PVC Jacket.

<u>Shielding</u>: Each pair: Aluminum-polyester shielding tape and a tinned copper drain wire. Aluminum foil surface facing in. Tape will be made with foil-free edges to ensure shield isolation or separate isolation tapes will be used. Overall: no overall shield.

<u>Jacket</u>: PVC. Thickness according to UL 444. Color: Gray unless specified.

<u>Diameter:</u> Nominal diameter shall not exceed that in table 324UH Column B.

<u>Flame test</u>: UL 1685 or higher.

<u>Voltage</u>: 300V but not printed on the jacket per UL 444 and UL 13.

<u>Application</u>: Communication Cable per NEC Art. 800 and UL 444; also for Class 3 or Class 2 Remote-Control, Signaling, and Power-Limited Circuits per NEC Art. 725.

<u>Installation</u>: Suitable for open cabling use in horizontal concealed spaces as CM, CL3, or CL2;
Not for use in risers or plenum spaces unless running in conduit. Items rated CMR, CL3R, or CL2R may be run in risers.

<u>Temp.</u>: 0C to 60C; or as indicated in Table 323UH. Consult manufacturers' specification.

<u>QA</u>: UL Listed as cable type CM. Cable ratings shall be printed on the jacket.
UL file number will be printed on the jacket alongside the gauge size and cable rating.
UL label shall be displayed on every cable package near manufacturer's label.

<u>Packaging</u>: 1000ft or 500ft: spools, reels or boxes.

<u>Appr. Manf</u>: Manufacturer make and model shall be as listed in table 324UH.

<u>Availability</u>: Cable shall be available from local distribution known to project management.

Table 324UH appears on page
13410-102 and 13410-103

24AWG cables are addressed in Public Cable 1.0 Specification 328C

Table 323UH **Stranded Tinned Copper; Shielded Pairs** **EPS/CD**
Str TC / Upgrade Insulation & Jacket / Shielded Pairs / no OA Shld (U) / Jacket **Horizontal Rated (H). some Riser Rated (R).**

Each Pair Shielded (EPS) / Common Drain Wire (CD)

Const	SDN	INS	JKT	CAP	TEMP	B	C	TAPPAN	GENERAL CABLE	BELDEN West Penn	BELDEN
22/2pr	323UH.222A	1	1	3	60C	.160	19	A29925	C1352A	77510	8723
	323UR.222B	1	1	3	75C	.160	19	R29821	-----	-----	-----
20/2pr	323UH.202A	3	1	4	80C	.305	47	-----	C7106A*	-----	-----
	323UR.202B	1	1	3	75C	.252	29	A39986	-----	-----	-----
18/2pr	323UR.182	1	1	3	75C	.267	36	R49893	-----	-----	-----

* Items with an asterisk also have an AWM rating.

Table 324UH **Stranded Tinned Copper; Shielded Pairs** **EPS**
Str TC / Upgrade Insulation & Jacket / Shielded Pairs / no OA Shld (U) /Jacket **Horizontal Rated (H). some Riser Rated (R).**

Each Pair Shielded (EPS) / Individual Drain Wires (1) per Pair

Const	SDN	INS	JKT	CAP	TEMP	B	C	TAPPAN	GENERAL CABLE	BELDEN West Penn	BELDEN
For 24 AWG RS-422 Cables see specification 328C											
22/2pr	324UH.222A	2	1	1	75C	.185	27	L29984	-----	-----	-----
	324UR.222C	1	1	3	75C	.160	22	R29822	-----	-----	-----
	324UH.222E	1	1	3	60C	.235	30	-----	-----	D430	-----
	324UH.222D	3	1	4	80C	.225	28	-----	C1350*	-----	-----
22/3pr	324UR.223C	1	1	2	80C	.255	43	R29320*	-----	-----	-----
	324UH.223D	1	1	3	80C	.292	44	SM2994*	C6040A*	-----	8777*
	324UH.223E	1	1	3	60C	.258	40	-----	-----	D431	-----
22/4pr	324UH.224A	2	1	1	75C	.290	50	L29982	-----	-----	-----
	324UH.224C	1	1	2	75C	.270	44	R29324	-----	-----	-----
	324UH.224E	1	1	3	60C	.280	49	-----	-----	D439	-----
	324UR.224F	3	1	4	80C	.373	72	R29331*	-----	-----	-----
22/6pr	324UH.226A	2	1	1	75C	.310	68	L29981	-----	-----	-----
	324UH.226B	1	1	2	75C	.300	69	SM2983	-----	-----	-----
	324UR.226C	1	1	2	80C	.300	69	R29323*	-----	-----	-----
	324UH.226D	1	1	3	80C	.381	83	SM2992*	C6041A*	-----	8778*
	324UH.226E	1	1	3	60C	.332	75	-----	-----	D432	-----
	324UH.226F	3	1	4	105C	.310	72	R29406	-----	-----	-----
22/8pr	324UR.228C	3	1	2	75C	.329	83	R29325	-----	-----	-----

* Items with an asterisk also have an AWM rating.

Manufacturer's makes vary. Diameters (Col. B) and weights (Col. C) are conservative and for engineering purposes only. Consult manufacturer's data for individual make & model information.

INS:	'1'=Polyethylene or Polypropylene, 2=Foam Polyethylene (FPE); 3=PVC
JKT:	'1'=PVC
CAP:	Capacitance in pF/ft '1' = <16 '2'= 16-24 '3'=25-35 '4'=36-49 '5'= 50 & up
Column B:	Cable Diameter in inches. Manufacturers' vary.
Column C:	Nominal cable weight in pounds per 1000ft

Table 324UH (continued) **Stranded Tinned Copper; Shielded Pairs** **EPS**
Str TC / Upgrade Insulation & Jacket / Shielded Pairs / no OA Shld (U) / Jacket **Horizontal Rated (H). some Riser Rated (R).**

Each Pair Shielded (EPS) / Individual Drain Wires (1) per Pair

Const	SDN	INS	JKT	CAP	TEMP	B	C	TAPPAN	GENERAL CABLE	BELDEN West Penn	BELDEN
22/9pr	324UR.229	1	1	2	75C	.349	92	R29346	-----	-----	-----
	324UH.229	1	1	3	80C	.445	113	SM2991*	C6042A*	-----	8774*
	324UH.229	1	1	3	60C	.390	94	-----	-----	D433	-----
22/11pr	324UR.2211	1	1	3	75C	.352	105	R29811	-----	-----	-----
	324UH.2211	1	1	3	80C	.486	130	SM2990*	C6043A*		8775*
22/12pr	324UR.2212	1	1	2	75C	.388	119	R29322	-----	-----	-----
	324UH.2212	1	1	3	80C	.533	150	SM2989*	C6059A*	-----	9768*
	324UH.2212	1	1	3	60C	.420	128	-----	-----	D434	-----
22/15pr	324UR.2215	1	1	2	75C	.418	146	R29344	-----	-----	-----
	324UH.2215	1	1	3	80C	.591	197	SM2988*	C6044A*	-----	8776*
22/17pr	324UH.2217	1	1	3	80C	.622	215	SM2987*	C6060A*		9769*
22/19pr	324UR.2219	1	1	2	75C	.447	181	R29343	-----	-----	-----
	324UH.2219	1	1	3	80C	.622	244	SM2986*	C6045A*	-----	8769*
	324UH.2219	1	1	3	60C	.565	185	-----	-----	D436	-----
22/20pr	324UR.2220	1	1	3	75C	.450	189	R29808	-----	-----	-----
22/25pr	324UR.2225	1	1	3	75C	.507	234	R29807	-----	-----	-----
22/27pr	324UR.2227	1	1	2	75C	.535	252	R29342	-----	-----	-----
	324UH.2227	1	1	3	80C	.709	341	SM2985*	C6046A*	-----	8773*
	324UH.2227	1	1	3	60C	.678	---	-----	-----	D438	-----
22/30pr	324UR.2230	1	1	3	75C	.544	281	R29806	-----	-----	-----
22/35pr	324UR.2235	1	1	3	75C	.575	324	R29805	-----	-----	-----
22/37pr	324UH.2237	1	1	3	80C	.800	481	SM2984*	-----	-----	9767*
18/2pr	324UH.182	1	1	3	80C	.334	56	SM4993*	-----	-----	-----
	324UR.182	1	1	3	75C	.267	40	R49676	-----	-----	-----
	324UH.182	1	1	4	60C	.268	47	-----	-----	D440	-----
18/3pr	324UH.183	1	1	3	80C	.404	83	SM4997*	C6047A*	-----	9773*
	324UR.183	1	1	3	75C	.286	55	R49890	-----	-----	-----
	324UH.183	1	1	4	60C	.285	65	-----	-----	D441	-----
18/4pr	324UR.184	1	1	3	75C	.336	72	R49889	-----	-----	-----
	324UH.184	1	1	4	60C	.312		-----	-----	D442	-----
	324UR.184	3	1	4	90C	.590	139	R49646*	-----	-----	-----
18/6pr	324UH.186	1	1	3	80C	.556	174	SM4989*	C6048A*	-----	9774*
	324UR.186	1	1	3	75C	.361	102	R49887	-----	-----	-----
18/9pr	324UH.189	1	1	3	80C	.655	247	SM4990*	C6049A*	-----	9775*
	324UR.189	1	1	3	75C	.424	148	R49884	-----	-----	-----
18/12pr	324UH.1812	1	1	3	80C	.735	309	SM4991*	C6050A*	-----	9776*
	324UR.1812	1	1	3	75C	.475	194	R49881	-----	-----	-----
18/15pr	324UH.1815	1	1	3	80C	.820	378	SM4992*	C6051A*	-----	9777*
	324UR.1815	1	1	3	75C	.513	239	R49880	------	-----	------

* Items with an asterisk also have an AWM rating.

Manufacturer's makes vary. Diameters (Col. B) and weights (Col. C) are conservative and for engineering purposes only. Consult manufacturer's data for individual make & model information.

Additional table notes on page 102 also apply to page 103.

323UP: 300V Specification 323: Stranded Tinned Copper; Shielded Pairs EPS/CD
Str TC / Upgrade Insulation & Jacket / Shielded Pairs / no OA Shld (U) / Jacket. Plenum Rated (P).

Each Pair Shielded (EPS) / Common Drain Wire (CD)

Conductor:	Stranded Tinned Copper
Insulation:	According to Table 323UP. Thickness according to UL 444.
Color Code:	Pair (1) Black& Red, (2) White & Green
Assembly:	Insulated conductors twisted into pairs and shielded (foil side out.). Multiple pairs cabled together with a common drain wire. Ripcord for jacket removal. PVC Jacket.
Shielding:	Each pair: Aluminum-polyester shielding tape. Aluminum foil surface facing out. Tinned Copper Drain Wire. Drain wire is outside the tapes and in continuous contact with the both foil shields.
Jacket:	Material according to Table 323UP. Thickness according to UL 444. Color: White or Red unless specified.
Diameter:	Nominal diameter shall not exceed that in table 323UP Column B.
Flame test:	NFPA 262 (Previously UL 910.)
Voltage:	300V but not printed on the jacket per UL 444.
Application:	Communication Cable per NEC Art. 800 and UL 444; also for Class 3 or Class 2 Remote-Control, Signaling, and Power-Limited Circuits per NEC Art. 725.
Installation:	Suitable for open cabling use in plenum spaces per the NEC as CMP, CL3P, or CL2P.
Temp.:	0C to 60C, or as indicated in Table 323UP. Consult manufacturer's specifications.
QA:	UL Listed as cable type CMP. Cable ratings shall be printed on the jacket. UL file number will be printed on the jacket alongside the gauge size and cable rating. UL label shall be displayed on every cable package near manufacturer's label.
Packaging:	1000ft or 500ft: spools, reels or boxes.
Appr. Manf:	Manufacturer make and model shall be as listed in table 323UP.
Availability:	Cable shall be available from local distribution known to project management.

Table 323UP **Stranded Tinned Copper; Shielded Pairs** **EPS/CD**
Str TC / Upgrade Insulation & Jacket / Shielded Pairs / no OA Shield (U) / Jkt. **Plenum Rated (P).**

Each Pair Shielded (EPS) / Common Drain Wire (CD) **Plenum**

Const	SDN	INS	JKT	CAP	TEMP	B	C	TAPPAN	GENERAL CABLE	BELDEN West Penn	BELDEN
22/2pr	323UP.222B	2	2	3	150C	.148	19	K29980	C3156	-----	87723
	323UP.222C	2	1	3	200C	.148	20	T29982	-----	-----	88723
	323UP.222D	2	3	4	75C	.153	18	M29978	-----	-----	82723
	323UP.222E	3	3	4	75C	.158	19	P29784	-----	-----	1325A
	323UP.222G	3	3	5	60C	.195	22	-----	-----	------	-----
18/2pr	323UP.182E	3	3	4	75C	.255	36	P49901	-----	-----	-----

Manufacturer's makes vary. Diameters (Col. B) and weights (Col. C) are conservative and for engineering purposes only. Consult manufacturer's data for individual make & model information.

INS: '1'=Foam FEP or Foam Fluoropolymer, '2'=FEP or other fluoropolymer, '3'=LS-PVC
JKT: '1'=FEP. '2'=other fluoropolymer. '3'= LS-PVC
CAP: Capacitance in pF/ft '1' = <16 '2'= 16-24 '3''=25-35 '4'=36-49 '5'= 50 & up
Column B: Cable Diameter in inches. Manufacturers' vary. Consult manufacturer's spec.'s.
Column C: Nominal cable weight in pounds per 1000ft

324UP: 300V Specification 324: Stranded Tinned Copper; Shielded Pairs EPS
Str TC / Upgrade Insulation & Jacket / Shielded Pairs / no OA Shield (U) / Jkt. Plenum Rated (P).
Each Pair Shielded (EPS) / Individual Drain Wires (1) per Pair

Conductor:	Stranded Tinned Copper. – 7 strands.
Insulation:	According to Table 324UP. Minimum thickness according to UL 444.
Color Code:	Two pair cable may be: Pair (1) Black& Red, (2) White&Green; or may follow as for three pairs and higher: Pair (1) Black& Red, (2) Black&White, (3) Black&Green, (4) Black&Blue, (5) Black&Yellow, (6) Black&Brown, (7) Black&Orange, (8) Red&White, (9) Red&Green. Etc. For color code above 9 pairs consult manufacturer's specification.
Assembly:	Insulated conductors twisted into pairs with a drain wire and tape shield (foil side in.) Multiple pairs cabled together. Shields are electrically isolated from each other. Ripcord for jacket removal. Jacket.
Shielding:	Each pair: Aluminum-polyester shielding tape and a tinned copper drain wire. Aluminum foil surface facing in. Tape will be made with foil-free edges to ensure shield isolation or separate isolation tapes will be used. Overall: no overall shield.
Jacket:	Material according to Table 324UP. Thickness according to UL 444. Color: White or Red unless specified.
Diameter:	Nominal diameter shall not exceed that in table 324UP Column B.
Flame test:	Passes NFPA 262 for use in plenum spaces.
Voltage:	300V but not printed on the jacket per UL 444.
Application:	Communication Cable per NEC Art. 800 and UL 444; also for Class 3 or Class 2 Remote-Control, Signaling, and Power-Limited Circuits per NEC Art. 725.
Installation:	Suitable for open cabling use in plenum spaces CMP, CL3P, or CL2P; also for open cabling in risers and in horizontal runs since rated higher than those requirements.
Temp.:	0C to 60C minimum range continuous operating ambient temperature.
QA:	UL Listed as cable type CMP. Cable ratings shall be printed on the jacket. UL file number will be printed on the jacket alongside the gauge size and cable rating. UL label shall be displayed on every cable package near manufacturer's label.
Packaging:	1000ft or 500ft: spools, reels or boxes.
Appr. Manf:	Manufacturer make and model shall be as listed in table 324UP.
Availability:	Cable shall be available from local distribution known to project management.

24AWG cables are addressed in Public Cable 1.0 Specification 328C

Table 324UP **Stranded Tinned Copper; Shielded Pairs** **EPS**
Str TC / Upgrade Insulation & Jacket / Shielded Pairs / no OA Shield (U) / Jkt. **Plenum Rated (P).**

Each Pair Shielded (EPS) / Individual Drain Wires (1) per Pair

Table 324UP Str. TC, Upgrade Insul.; EPS; no O/A shield, Plenum Rated

Const	SDN	INS	JKT	CAP	TEMP	B	C	TAPPAN	GENERAL CABLE	BELDEN West Penn	BELDEN
For 24 AWG RS-422 Cables see specification 328C											
22/2pr	324UP.222B	2	2	2	150C	.160	25	K29995	-----	-----	-----
	324UP.222C	2	1	3	200C	.186	26	-----	C8112	-----	-----
	324UP.222D	2	3	3	60C	.148	22	M29980	C8105	-----	-----
	324UP.222D	2	3	3	60C	.218	26	-----	-----	D25430	-----
22/3pr	324UP.223A	1	3	1	75C	.258	41	M29941	-----	-----	-----
	324UP.223B	2	2	3	150C	.234	40	K29979	C3157	-----	87777
	324UP.223C	2	1	3	200C	.234	41	T29978	-----	-----	88777
	324UP.223D	2	3	3	60C	.237	39	M29940	C8131	D25431	82777
	324UP.223E	3	3	4	75C	.204	32	P29811	-----	-----	1326A
22/4pr	324UP.224B	2	2	3	150C	.222	40	K29977	----	-----	-----
	324UP.224D	2	3	3	60C	.260	49	M29938	-----	D25439	-----
	324UP.224E	3	3	4	75C	.240	42	P29810	-----	-----	-----
22/6pr	324UP.226B	2	2	2-3	150C	.280	70	K29978	-----	-----	87778
	324UP.226C	2	1	3	200C	.309	75	T29981	C8132	-----	88778
	324UP.226D	2	3	3	60C	.314	75	M29937	C8133	D25432	82778
	324UP.226E	3	3	4	75C	.278	65	P29758	-----	-----	-----
22/8pr	324UP.228D	2	3	3	75C	.309	91	M29936	6608HS	-----	-----
	324UP.228E	3	3	4	75C	.285	78	P29756	-----	-----	-----
22/9pr	324UP.229E	3	3	4	75C	.307	89	P29755	-----	-----	-----
22/10pr	324UP.2210	3	3	4	75C	.329	99	P29754	-----	-----	-----
22/12pr	324UR.2212D	2	3	2	75C	.424	152	M29991	6612HS	-----	-----
	324UP.2212E	3	3	4	75C	.343	116	P29752	-----	-----	-----
22/15pr	324UP.2215E	3	3	4	75C	.370	143	P29751	-----	-----	-----
18/2pr	324UP.182E	3	3	5	75C	.255	40	P49898	-----	-----	-----
18/3pr	324UP.183C	2	1	3	200C	.250	54	T49984	-----	-----	-----
	324UP.183E	3	3	5	75C	.274	56	P49897	-----	-----	-----

Manufacturer's makes vary. Diameters (Col. B) and weights (Col. C) are conservative and for engineering purposes only. Consult manufacturer's data for individual make & model information.

INS:	'1'=Foam FEP or Foam Fluoropolymer, '2'=FEP or other fluoropolymer, '3'=LS-PVC
JKT:	'1'=FEP. '2'=other fluoropolymer. '3'= LS-PVC
CAP:	Capacitance in pF/ft '1' = <16 '2'= 16-24 '3"=25-35 '4'=36-49 '5'= 50 & up
Column B:	Cable Diameter in inches. Manufacturers' vary. Consult manufacturer's spec.'s.
Column C:	Nominal cable weight in pounds per 1000ft

328AH: 300V Specification 328A: Paired Str. TC; Double O/A Shield **EIA-485 / RS-485**
Stranded TC / Upgrade Insul. 120Ω / Pairs / Double OAS (Foil+90% Braid) / Jacket. **Horizontal Rated**

Conductor:	24 AWG Stranded Tinned Copper .
Insulation:	Polyethylene, thickness as required for 120 Ohm Impedance. Minimum thickness according to UL 444.
Color Code:	Pair (1) White w/Blue Stripe & Blue w/White Stripe; Pair (2) White w/Orange Stripe & Orange w/White Stripe Pair (3) White w/Green Stripe & Green w/White Stripe Pair (4) White w/Brown Stripe & Brown w/White Stripe.
Assembly:	Twisted pairs; no individual pair shielding. Multiple pairs cabled together. Overall shielding..
Shielding:	Overall double shield of aluminum-polyester tape with a drain wire; plus 90% tinned copper braid.
Jacket:	PVC. Thickness according to UL 444 and applicable AWM rating if any. Color: Gray unless specified.
Diameter:	Nominal diameter shall not exceed that in table 328AH Column B.
Flame test:	Passes UL 1685 Vertical Tray (UL 1581).
Voltage:	300V but not printed on the jacket per UL 444.
Application:	Communication Cable per NEC Art. 800 and UL 444. Designed for use interconnecting EIA-485 devices..
Installation:	Suitable for open cabling use in horizontal runs as Type CM – communications cable. Not for use in risers or plenum spaces unless running in conduit.
Temp.:	0C to 60C unless the jacket is printed with ratings outside this range; as for use as appliance wiring material type AWM/
QA:	UL Listed as cable type CM. Cable ratings shall be printed on the jacket. UL file number will be printed on the jacket alongside the gauge size and cable rating. UL label shall be displayed on every cable package near manufacturer's label.
Packaging:	1000ft or 500ft: spools, reels or boxes.
Appr. Manf:	Manufacturer make and model shall be as listed in table 328AH.
Availability:	Cable shall be available from local distribution known to project management.

Table 328AH **Str. TC, Upgrade Insul.; Pair(s), Double Overall Shield, Horizontal Rated 120 Ohm Nom. Impedance; Overall Braid is 90% EIA-485 design purpose**

Const	SDN	INS	JKT	CAP	TEMP	B	C	TAPPAN	GENERAL CABLE	BELDEN West Penn	BELDEN
24/1pr	328AH.241A	1	1	1	80C	.236	40	L19827*	C0841A*	-----	9841*
	328AH.241B	1	1	1	60C	.275	36	-----	-----	D4851	-----
24/2pr	328AH.242A	1	1	1	80C	.353	58	L19954*	C0842A*	-----	9842*
	328AH.242B	1	1	1	60C	.305	62	-----	-----	D4852	-----
24/3pr	328AH.243A	1	1	1	80C	.360	67	L19773*	C0843A*	-----	9843*
	328AH.243B	1	1	1	60C	.320	73	-----	-----	D4853	-----
24/4/pr	328AH.244A	1	1	1	80C	.390	83	L19893*	C0844A*	-----	9844*
	328AH.244B	1	1	1	60C	.345	81	-----	-----	D4854	-----

* Items with an asterisk also have an AWM rating.

Manufacturer's makes vary. Diameters (Col. B) and weights (Col. C) are conservative and for engineering purposes only. Consult manufacturer's data for individual make & model information.

INS: '1'=Polyethylene or Polypropylene, 2=Foam Polyethylene (FPE); 3=PVC
JKT: '1'=PVC
CAP: Capacitance in pF/ft '1' = <16
Column B: Cable Diameter in inches. Manufacturers' vary. Consult manufacturer's spec.'s.
Column C: Nominal cable weight in pounds per 1000ft

328AP: 300V Specification 328A: Paired Str. TC; Double O/A Shield **EIA-485 / RS-485**
Stranded TC / Upgrade Insul. 120Ω / Pairs / Double OAS (Foil+90% Braid) / Jacket. **Plenum Rated**

Conductor: 24 AWG Stranded Tinned Copper.

Insulation: Foam FEP or Foam Fluoropolymer; thickness as required for 120 Ohm Impedance.
Minimum thickness according to UL 444.

Color Code: Pair (1) White w/Blue Stripe & Blue w/White Stripe;
Pair (2) White w/Orange Stripe & Orange w/White Stripe
Pair (3) White w/Green Stripe & Green w/White Stripe
Pair (4) White w/Brown Stripe & Brown w/White Stripe.

Assembly: Twisted pairs; no individual pair shielding. Multiple pairs cabled together. Overall shielding.

Shielding: Overall double shield of aluminum-polyester tape with a drain wire;
plus 90% tinned copper braid.

Jacket: Material according to table 328AP or manufacturer's specification.
Thickness according to UL 444 and applicable AWM rating if any.
Color: White or natural unless specified.

Diameter: Nominal diameter shall not exceed that in table 328AP Column B.

Flame test: Passes NFPA 262.

Voltage: 300V but not printed on the jacket per UL 444.

Application: Communication Cable per NEC Art. 800 and UL 444. Designed for use interconnecting EIA-485 devices..

Installation: Suitable for open cabling use in plenum spaces as Type CMP. Also for use also as open cabling in risers and in horizontal runs since rated higher those requirements.

Temp.: 0C minimum. Upper temperature rating.according to table 328AP; minimum range continuous operating ambient temperature.

QA: UL Listed as cable type CMP. Cable ratings shall be printed on the jacket.
UL file number will be printed on the jacket alongside the gauge size and cable rating.
UL label shall be displayed on every cable package near manufacturer's label.

Packaging: 1000ft or 500ft: spools, reels or boxes.

Appr. Manf: Manufacturer make and model shall be as listed in table 328AP.

Availability: Cable shall be available from local distribution known to project management.

Table 328AP **Str. TC, Upgrade Insul.; Pair(s), Double Overall Shield, Plenum Rated 120 Ohm Nom. Impedance; Overall Braid is 90% EIA-485 design purpose**

Const	SDN	INS	JKT	CAP	TEMP	B	C	TAPPAN	GENERAL CABLE	BELDEN West Penn	BELDEN
24/1pr	328AP.241A	1	3	1	75C	.204	26	M19899	-----	D254851	82841
	328AP.241B	1	1	1	200C	.202	27	T19984	C8117	-----	89841
24/2pr	328AP.242A	1	3	1	75C	.305	42	M19896	C8129	D254852	82842
	328AP.242B	1	1	1	200C	.305	49	T19983	-----	-----	89842
24/4pr	328AP.244	1	2	1	150C	.290	51	K19941	-----	-----	-----

Manufacturer's makes vary. Diameters (Col. B) and weights (Col. C) are conservative and for engineering purposes only. Consult manufacturer's data for individual make & model information.

INS: '1'=Foam FEP or Foam Fluoropolymer, '2'=FEP or other fluoropolymer, '3'=LS-PVC
JKT: '1'=FEP. '2'=other fluoropolymer. '3'= LS-PVC
CAP: Capacitance in pF/ft '1' = <16
Column B: Cable Diameter in inches. Manufacturers' vary. Consult manufacturer's spec.'s.
Column C: Nominal cable weight in pounds per 1000ft

328BH: 300V Specification 328B: Paired Str. TC; Double O/A Shield
Stranded TC / Upgrade Insul. 100Ω / Pairs / Double OAS (Foil+60% Braid) / Jacket.

EIA-232 / RS-232
EIA-422 / RS-422
Horizontal Rated

Conductor:	Stranded Tinned Copper .
Insulation:	Material according to Table 328BH, thickness as required for 100 Ohm Impedance. Minimum thickness according to UL 444 or AWM Style as applicable.
Color Code:	Pair (1) White w/Blue Stripe & Blue w/White Stripe; Pair (2) White w/Orange Stripe & Orange w/White Stripe Pair (3) White w/Green Stripe & Green w/White Stripe Pair (4) White w/Brown Stripe & Brown w/White Stripe. (etc.) Pair (25) Violet w/Gray Stripe & Gray w/Violet Stripe.
Assembly:	Twisted pairs; no individual pair shielding. Multiple pairs cabled together. Overall shielding.
Shielding:	Overall double shield of aluminum-polyester tape with a drain wire; plus 60% tinned copper braid.
Jacket:	PVC. Thickness according to UL 444 and applicable AWM rating if any. Color: Gray unless specified.
Diameter:	Nominal diameter shall not exceed that in table 328BH Column B.
Flame test:	Passes UL 1685 Vertical Tray (UL 1581).
Voltage:	300V but not printed on the jacket per UL 444.
Application:	Communication Cable per NEC Art. 800 and UL 444. Designed for use interconnecting EIA-232 devices (RS-232) or EIA-422 devices (RS-422.)
Installation:	Suitable for open cabling use in horizontal runs as Type CM communications cable. Not for use in risers or plenum spaces unless running in conduit.
Temp.:	0C to 80C minimum range continuous operating ambient temperature.
QA:	UL Listed as cable type CM. Cable ratings shall be printed on the jacket. UL file number will be printed on the jacket alongside the gauge size and cable rating. UL label shall be displayed on every cable package near manufacturer's label.
Packaging:	1000ft or 500ft: spools, reels or boxes.
Appr. Manf:	Manufacturer make and model shall be as listed in table 328BH.
Availability:	Cable shall be available from local distribution known to project management.

Table 328BH **Str. TC, Upgrade Insul.; Pair(s), Double Overall Shield, Horizontal Rated**
AWM Rated **100 Ohm Nom. Imp.; Overall Braid is 60% EIA-232 and EIA-422 design purpose**

Const	SDN	INS	JKT	CAP	TEMP	B	C	TAPPAN	GENERAL CABLE	BELDEN
24/2pr	328BH.242A	2	1	1	80C	.276	38	L19772	C0515A	8102
	328BH.242B	1	1	1	80C	.291	43	L19909	C0829A	9829
	328BH.242C	3	1	3	80C	.250	37	R19768	C0620A	8332
24/3pr	328BH.243A	2	1	1	80C	.290	42	L19853	C0516A	8103
	328BH.243B	1	1	1	80C	.305	53	L19908	C0830A	9830
	328BH.243C	3	1	3	80C	.265	44	R19767	C0621A	8333
24/4pr	328BH.244A	2	1	1	80C	.315	48	L19919	C0517A	8104
	328BH.244B	1	1	1	80C	.330	58	L19892	C0831A	9831
	328BH.244C	3	1	3	80C	.288	49	R19766	C0622A	8334
24/5pr	328BH.245A	2	1	1	80C	.340	55	L19918	C0518A	8105
	328BH.245B	1	1	1	80C	.338	65	L19907	C0832A	9832
	328BH.245C	3	1	3	80C	.295	57	R19765	C0623A	8335
24/6pr	328BH.246A	2	1	1	80C	.368	62	L19917	C0519A	8106
	328BH.246B	1	1	1	80C	.364	69	L19906	C0839A	9839
	328BH.244C	3	1	3	80C	.311	62	R19764	C0624A	8336
24/7pr	328BH.247A	2	1	1	80C	.370	67	L19916	C0520A	8107
	328BH.247B	1	1	1	80C	.370	77	L19905	C0833A	9833
	328BH.247C	3	1	3	80C	.321	65	R19763	C0625A	8337
24/8pr	328BH.248A	2	1	1	80C	.397	75	L19915	C0521A	8108
	328BH.248C	3	1	1	80C	.336	76	-----	C0626A	-----
24/9pr	328BH.249B	1	1	1	80C	.419	93	L19904	-----	9834
24/10pr	328BH.2410A	2	1	1	80C	.473	90	L19914	C0522A	8110
	328BH.2410B	1	1	1	80C	.451	102	L19903	C0835A	9835
	328BH.2410C	3	1	3	80C	.385	90	R19762	C6028A	8340
24/12pr	328BH.2412B	1	1	1	80C	.464	114	L19902	C0836A	9836
24/12½pr	328BH.2412.5A	2	1	1	80C	.486	111	L19913	C0523A	8112
	328BH.2412.5C	3	1	3	80C	.405	109	R19761	C0630A	8342
24/15pr	328BH.2415A	2	1	1	80C	.555	127	L19912	C0524A	8115
	328BH.2415C	3	1	3	80C	.445	123	R19760	-----	8345
24/18pr	328BH.2418A	2	1	1	80C	.585	150	L19911	C0525A	8118
	328BH.2418B	1	1	1	80C	.567	174	L19901	-----	9837
	328BH.2418C	3	1	3	80C	.480	152	R19759	-----	8348
24/25pr	328BH.2425A	2	1	1	80C	.677	201	L19910	C0526A	8125
	328BH.2425B	1	1	1	80C	.670	226	L19900	-----	9838
	328BH.2425C	3	1	3	80C	.550	195	R19758	-----	8355

Manufacturer's makes vary. Diameters (Col. B) and weights (Col. C) are conservative and for engineering purposes only. Consult manufacturer's data for individual make & model information.

INS:	'1'=Polyethylene or Polypropylene, 2=Foam Polyethylene (FPE); 3=PVC
JKT:	'1'=PVC
CAP:	Capacitance in pF/ft '1' = <16 '2'= 16-24 '3''=25-35
Column B:	Cable Diameter in inches. Manufacturers' vary. Consult manufacturer's spec.'s.
Column C:	Nominal cable weight in pounds per 1000ft

328BP: 300V Specification 328B: Paired Str. TC; Double O/A Shield
Stranded TC / Upgrade Insul. 100Ω / Pairs / Double OAS (Foil+60% Braid) / Jacket.

EIA-232 / RS-232
EIA-422 / RS-422
Plenum Rated

Conductor: 24 AWG Stranded Tinned Copper.

Insulation: Foam FEP or Foam Fluoropolymer; thickness as required for 100 Ohm Impedance. Minimum thickness according to UL 444.

Color Code: Pair (1) White w/Blue Stripe & Blue w/White Stripe;
Pair (2) White w/Orange Stripe & Orange w/White Stripe
Pair (3) White w/Green Stripe & Green w/White Stripe
Pair (4) White w/Brown Stripe & Brown w/White Stripe
(etc.)
Pair (25) Violet w/Gray Stripe & Gray w/Violet Stripe.

Assembly: Twisted pairs; no individual pair shielding. Multiple pairs cabled together. Overall shielding.

Shielding: Overall double shield of aluminum-polyester tape with a drain wire; plus 60% tinned copper braid.

Jacket: Material according to table 328BP or manufacturer's specification. Thickness according to UL 444. Color: White or natural unless specified.

Diameter: Nominal diameter shall not exceed that in table 328BP Column B.

Flame test: Passes NFPA 262.

Voltage: 300V but not printed on the jacket per UL 444.

Application: Communication Cable per NEC Art. 800 and UL 444. Designed for use interconnecting EIA-233 and EIA-422 devices..

Installation: Suitable for open cabling use in plenum spaces as Type CMP. Also for use also as open cabling in risers and in horizontal runs since rated higher those requirements.

Temp.: 0C minimum. Upper temperature rating according to table 328BP; minimum range continuous operating ambient temperature.

QA: UL Listed as cable type CMP. Cable ratings shall be printed on the jacket.
UL file number will be printed on the jacket alongside the gauge size and cable rating.
UL label shall be displayed on every cable package near manufacturer's label.

Packaging: 1000ft or 500ft: spools or reels.

Appr. Manf: Manufacturer make and model shall be as listed in table 328BP.

Availability: Cable shall be available from local distribution known to project management.

Table 328BP **Str. TC, Upgrade Insul.; Pair(s), Double Overall Shield, Plenum Rated 100 Ohm Nom. Imp.; Overall Braid is 60%; EIA-232 and EIA-422 design purpose**

Const	SDN	INS	JKT	CAP	TEMP	B	C	TAPPAN	BELDEN
24/2pr	328BP.242	1	2	1	150C	.216	20	K19970	88102
24/3pr	328BP.243	1	2	1	150C	.239	31	K19955	88103
24/4pr	328BP.244	1	2	1	150C	.259	38	K19954	88104
24/5pr	328BP.245	1	2	1	150C	.274	44	K19953	88105
24/6pr	328BP.246	1	2	1	150C	.295	50	K19952	88106
24/7½pr	328BP.247.5	1	2	1	150C	.327	59	K19951	88107
24/9pr	328BP.249	1	2	1	150C	.352	74	K19950	88109
24/12½pr	328BP.2412.5	1	2	1	150C	.397	97	-----	88112
24/18pr	328BP.2418	1	2	1	150C	.482	148	-----	88118
24/25pr	328BP.2425	1	2	1	150C	.581	195	-----	88125

Manufacturer's makes vary. Diameters (Col. B) and weights (Col. C) are conservative and for engineering purposes only. Consult manufacturer's data for individual make & model information.

INS: '1'=Foam FEP or Foam Fluoropolymer, '2'=FEP or other fluoropolymer, '3'=LS-PVC
JKT: '1'=FEP. '2'=other fluoropolymer. '3'= LS-PVC
CAP: Capacitance in pF/ft '1' = <16
Column B: Cable Diameter in inches. Manufacturers' vary. Consult manufacturer's spec.'s.
Column C: Nominal cable weight in pounds per 1000ft

328CH: 300V Specification 328C: Paired Str. TC; EPS, no O/A Shield
Stranded TC / Upgrade Insul. / Multiple Pairs / Each Pair Shielded

EIA-422 / RS-422
Horizontal Rated

Conductor: Stranded Tinned Copper .

Insulation: Material according to Table 328CH, thickness as required for 100 Ohm Impedance except 1 pair cables shall have insulation thickness according to manufacturer's specification. Minimum thickness according to UL 444 or AWM Style as applicable.

Color Code: 1 pair cables: Black & Clear. Multipair cables:
Pr 1 Black & Red; Pr 2 Black & White; Pr 3 Black & Green; Pr 4 Black & Blue;
Pr 5 Black & Yellow; Pr 6 Black & Brown; Pr 7 Black & Orange; Pr 8 Red & White;
Pr 9 Red & Green; Pr 10 Red & Blue; Pr 11 Red & Yellow; Pr 12 Red & Brown;
Pr 13 Red & Orange; Pr 14 Green & White; Pr 15 Green & Blue; Pr 16 Green & Yellow;
Pr 17 Green & Brown; Pr 18 Green & Orange; Pr 19 White & Blue; Pr 20 White & Yellow;
Pr 21 White & Brown; Pr 22 White & Orange; Pr 23 Blue & Yellow; Pr 24 Blue & Brown;
Pr 25 Blue & Orange; Pr 26 Brown & Yellow; Pr 27 Brown & Orange

Assembly: Twisted pairs; individually shielded. Multiple pairs cabled together. Jacketed.

Shielding: Pairs individually shielded with aluminum-polyester tape with a drain wire.

Jacket: PVC. Thickness according to UL 444 and applicable AWM rating if any. Color: Gray unless specified.

Diameter: Nominal diameter shall not exceed that in table 328CH Column B.

Flame test: Passes UL 1685 Vertical Tray (UL 1581).

Voltage: 300V but not printed on the jacket per UL 444.

Application: Communication Cable per NEC Art. 800 and UL 444. Designed for use interconnecting EIA-422 devices (RS-422.)

Installation: Suitable for open cabling use in horizontal runs as Type CM communications cable. Not for use in risers or plenum spaces unless running in conduit.

Temp.: 0C to 60C minimum range, continuous operating ambient temperature.

QA: UL Listed as cable type CM. Cable ratings shall be printed on the jacket. UL file number will be printed on the jacket alongside the gauge size and cable rating. UL label shall be displayed on every cable package near manufacturer's label.

Packaging: 1000ft or 500ft: spools, reels or boxes.

Appr. Manf: Manufacturer make and model shall be as listed in table 328CH.

Availability: Cable shall be available from local distribution known to project management.

Table 328CH Str. TC, Upgrade Insul.; 1 Pair Shielded, Horizontal Rated

Const	SDN	INS	JKT	CAP	TEMP	B	C	TAPPAN	GENERAL CABLE	BELDEN
24/1pr	328CH.241	2	1	1	60	.150	12	L19869	-----	-----
	328CH.241	1	1	2	60	.168	14	L10051*	C2513A*	8641*

* Items with an asterisk are also AWM rated.

Table 328CH Str. TC, Upgrade Insul.; Pairs EPS, no O/A Shield, Horizontal Rated
AWM Rated EIA-422 design purpose

Const	SDN	INS	JKT	CAP	TEMP	B	C	TAPPAN	GENERAL CABLE	BELDEN
24/2pr	328CH.242	2	1	1	60	.340	45	L19887	C0910A	9729
24/3pr	328CH.243	2	1	1	60	.381	54	L19886	C0911A	9730
24/4pr	328CH.244	2	1	1	60	.416	65	L19888	C0912A	9728
24/6pr	328CH.246	2	1	1	60	.492	92	L19885	C0913A	9731
24/9pr	328CH.249	2	1	1	60	.601	124	L19884	C0914A	9732
24/11pr	328CH.2411	2	1	1	60	.652	156	L19883	C0915A	9733
24/12pr	328CH.2412	2	1	1	60	.672	166	L19882	C0916A	9734
24/15pr	328CH.2415	2	1	1	60	.743	196	L19881	C0917A	9735
24/17pr	328CH.2417	2	1	1	60	.671	216	L19880	-----	9736
24/19pr	328CH.2418	2	1	1	60	.671	234	L19879	-----	9737
24/27pr	328CH.2427	2	1	1	60	.797	334	L19878	-----	9738

Manufacturer's makes vary. Diameters (Col. B) and weights (Col. C) are conservative and for engineering purposes only. Consult manufacturer's data for individual make & model information.

INS:	'1'=Polyethylene or Polypropylene, 2=Foam Polyethylene (FPE); 3=PVC
JKT:	'1'=PVC
CAP:	Capacitance in pF/ft '1' = <16 '2'= 16-24 '3"=25-35
Column B:	Cable Diameter in inches. Manufacturers' vary. Consult manufacturer's spec.'s.
Column C:	Nominal cable weight in pounds per 1000ft

328CP: 300V Specification 328C: Paired Str. TC; EPS, no O/A Shield
Stranded TC / Upgrade Insul. / Multiple Pairs / Each Pair Shielded / Jacket

EIA-422 / RS-422
EIA-232 / RS-232
Plenum Rated

Conductor: Stranded Tinned Copper .

Insulation: Material according to Table 328CP, thickness as required for 100 Ohm Impedance except 1 pair cables shall have insulation thickness according to manufacturer's specification. Minimum thickness according to UL 444.

Color Code: Pair (1) White w/Blue Stripe & Blue w/White Stripe;
Pair (2) White w/Orange Stripe & Orange w/White Stripe
Pair (3) White w/Green Stripe & Green w/White Stripe
Pair (4) White w/Brown Stripe & Brown w/White Stripe
(etc.)
Pair (18) Yellow w/Green Stripe & Green w/Yellow Stripe

Assembly: Twisted pairs; individually shielded. Multiple pairs cabled together. Jacketed.

Shielding: Pairs individually shielded with aluminum-polyester tape with a drain wire.

Jacket: Material according to Table 328C. Thickness according to UL 444.

Diameter: Nominal diameter shall not exceed that in table 328CP Column B.

Flame test: Passes NFPA 262 (formerly UL 910).

Voltage: 300V but not printed on the jacket per UL 444.

Application: Communication Cable per NEC Art. 800 and UL 444. Multi-pair cables are designed for use interconnecting EIA-422 devices (RS-422.) Single pair cables are designed for use interconnecting EIA-232 devices (RS-232.)

Installation: Suitable for open cabling use in plenum spaces Type CMP communications cable. Also suitable for use risers and horizontal runs since rated higher. Suitable for use in conduit.

Temp.: 0C to 60C minimum range, continuous operating ambient temperature.

QA: UL Listed as cable type CMP. Cable ratings shall be printed on the jacket.
UL file number will be printed on the jacket alongside the gauge size and cable rating.
UL label shall be displayed on every cable package near manufacturer's label.

Packaging: 1000ft or 500ft: spools, reels or boxes.

Appr. Manf: Manufacturer make and model shall be as listed in table 328CP.

Availability: Cable shall be available from local distribution known to project management.

Table 328CP **Str. TC, Upgrade Insul.; 1 OAS Pair, Plenum Rated**

EIA-232 design purpose

Const	SDN	INS	JKT	CAP	TEMP	B	C	TAPPAN	GENERAL CABLE	BELDEN
24/1pr	328CP.241A	2	3	3	75C	.108	10	M19871	C8127	82641
	328CP.241B	2	1	3	200	.106	10	T19977	-----	88641

Manufacturer's makes vary. Diameters (Col. B) and weights (Col. C) are conservative and for engineering purposes only. Consult manufacturer's data for individual make & model information.

INS: '1'=Foam FEP or Foam Fluoropolymer, '2'=FEP or other fluoropolymer, '3'=LS-PVC
JKT: '1'=FEP. '2'=other fluoropolymer. '3'= LS-PVC
CAP: Capacitance in pF/ft '1' = <16 '2'= 16-24 '3''=25-35 '4'=36-49 '5'= 50 & up
Column B: Cable Diameter in inches. Manufacturers' vary. Consult manufacturer's spec.'s.
Column C: Nominal cable weight in pounds per 1000ft

Table 328CP **Str. TC, Upgrade Insul.; Pair(s), Plenum Rated**

EIA-422 design purpose

Const	SDN	INS	JKT	CAP	TEMP	B	C	TAPPAN	BELDEN
24/2pr	328CP.242A	1	2	1	150C	.261	31	K19957	89729
	328CP.242B	1	3	1	75C	.255	28	M19923	82729
24/3pr	328CP.243	1	2	1	150C	.278	40	K19958	89730
24/4pr	328CP.244	1	2	1	150C	.307	50	K19990	89728
24/6pr	328CP.246	1	2	1	150C	.361	71	K19960	89731
24/9pr	328CP.249	1	2	1	150C	.429	108	K19964	89732
24/18pr	328CP.2418	1	2	1	150C	.459	155	K19962	-----

Manufacturer's makes vary. Diameters (Col. B) and weights (Col. C) are conservative and for engineering purposes only. Consult manufacturer's data for individual make & model information.

INS: '1'=Foam FEP or Foam Fluoropolymer, '2'=FEP or other fluoropolymer, '3'=LS-PVC
JKT: '1'=FEP. '2'=other fluoropolymer. '3'= LS-PVC
CAP: Capacitance in pF/ft '1' = <16
Column B: Cable Diameter in inches. Manufacturers' vary. Consult manufacturer's spec.'s.
Column C: Nominal cable weight in pounds per 1000ft

MASTER TABLE OF PUBLIC CABLE 1.0 PRODUCTS

PERIODIC TABLE OF PARTIAL CONTENTS – Detail for Category Cable

Category 3

Specification 333

Cable Type	Installation	Spec / Table	Spec. Page	Table Page
Category 3 UTP	Horiz. / Riser	333UH	122	122
	Plenum	333UP	123	123

Category 5e

Specification 335

Cable Type	Installation	Spec / Table	Spec. Page	Table Page
Category 5e UTP 350MHz	Horiz. / Riser	335UH	124	124
	Plenum	335UP	125	125

Category 6

Specification 336

Cable Type	Installation	Spec / Table	Spec. Page	Table Page
Category 6 UTP 250MHz	Horiz. / Riser	336UH	126	126
	Plenum	336UP	127	127

PERIODIC TABLE OF PARTIAL CONTENTS – Detail for Fire Detection and Alarm Cable

Cabled; Solid Bare Copper Fire Alarm Cable

Specification 341

Shielding	Installation	Spec / Table	Spec. Page	Table Page
Unshielded	Horiz. / Riser	341UH	128	129
	Plenum	341UP	130	131
Shielded	Horiz. / Riser	341SH	132	133
	Plenum	341SP	134	135

Cabled; Stranded Bare Copper Fire Alarm Cable

Specification 342

Shielding	Installation	Spec / Table	Spec. Page	Table Page
Unshielded	Riser	342UR	136	136
	Plenum	342UP	137	137
Shielded	Riser	342SR	138	138
	Plenum	342SP	139	139

NYC LL5 Fire Alarm Cable – 150C

Specification 343

Local Code & Temp	Shielding	Spec / Table	Spec. Page	Table Page
NYC LL5 150C	Unshielded	343UP	140	140
	Shielded	343SP	141	141

NYC LL5 Fire Alarm Cable – 200C

Specification 344

Local Code & Temp.	Shielding	Spec / Table	Spec. Page	Table Page
NYC LL5 200C	Unshielded	344UP	142	142
	Shielded	344SP	143	143

Fire Resistive Cables and Cable for use in Cable-In-Conduit Systems

Specification 345

Installation	Shielding	Spec / Table	Spec. Page	Table Page
Circuit Integrity FPLR-CI Open Wiring	Unshielded	345UCI	144	144
	Shielded	345SCI	145	145
Fire Resistive System Cable –In—Conduit / CIC	Unshielded	345UCIC	146	146
	Shielded	345SCIC	147	147

333UH: 300V Specification 333: Paired Solid BC **Category 3**
Solid bare copper / 100Ω Multiple Pairs / Unshielded (U) / Category 3 **Horizontal Rated (H).**

Conductor: 24 AWG Solid Bare Copper.

Insulation: Thickness according to UL 444 and as required for electrical characteristics. Material as required for flame test and for electrical characteristics.

Color Code: Per TIA 568 for Category 3 cable, and manufacturer's specification.

Assembly: Insulated conductors twisted into pairs. Multiple pairs cabled together. Unshielded Ripcord under the jacket to facilitate removal.

Jacket: PVC. Thickness according to UL 444. Color: Gray unless specified.

Diameter: Nominal diameter shall not exceed that in table 333UH Column B.

Flame test: Passes UL 1685 Vertical Tray (UL 1581) for CM rated models. Passes UL 1666 for CMR rated models.

Voltage: 300V but not printed on the jacket per UL 444.

Electricals: EIA-568 Category 3

Application: Communication Cable per NEC Art. 800 and UL 444. EIA 568 Category 3.

Installation: Models rated CMR are suitable for use in risers. Run as open cabling. Models rated CM are suitable for use as open cabling in horizontal runs. Not for use in plenum spaces unless running in conduit.

Temp.: 0C to 60C minimum range, continuous operating ambient temperature.

QA Flame: UL Listed as cable type CM or CMR. Cable ratings shall be printed on the jacket. UL file number will be printed on the jacket alongside the gauge size and cable rating. UL label shall be displayed on every cable package near manufacturer's label.

QA-Elec: NRTL certified Cat 3 (preferred.) Manufacturer self-certified Cat 3 accepted.

Packaging: 1000ft or 500ft: spools, reels or boxes.

Appr. Manf: Manufacturer make and model shall be as listed in table 333UH.

Availability: Cable shall be available from local distribution known to project management.

333UH: 300V Specification 333: Paired Solid BC **Category 3**

Const	SDN	B	C	TAPPAN	GENERAL CABLE	BELDEN-WEST PENN	BELDEN
24/2pr	333UH.2	.157	12	I19998.3	2133009	-----	1227A1
24/3pr	333UH.3	.168	16	I19988.3	2133013	-----	24501934
24/4pr	333UH.4	.191	21	I19990.3	2133017	-----	1229A1
24/6pr	333UH.6	.219	29	I19997.3	2133021	-----	24501829
24/12pr	333UH,12	.280	52	I19995.3	2133027	-----	24501837
24/25pr	333UH.25	.420	105	I19986.3	2133033	WP55700	1232A1
24/50pr	333UH.50	.560	219	I90001.3	2133161	WP55216	24501877
24/75pr	333UH.75	.630	315	-----	-----	-----	24501887
24/100pr	333UH.100	.770	393	I90002.3	2133144	WP55211	24501897
24/200pr	333UH.200	1.080	744	I90003.3	2133323	WP55212	24501906
24/300pr	333UH.300	1.230	1077	-----	2133373	-----	------

Manufacturer's makes vary. Diameters (Col. B) and weights (Col. C) are conservative and for engineering purposes only. Consult manufacturer's data for individual make & model information.

Column B: Nominal Cable Diameter in inches, for conduit fill calculations.
Column C: Nominal cable weight in pounds per 1000ft.

333UP: 300V Specification 333: Paired Solid BC **Category 3**
Solid bare copper / 100Ω Multiple Pairs / Unshielded (U) / Category 3 **Plenum Rated (P).**

Conductor: 24 AWG Solid Bare Copper.

Insulation: Thickness according to UL 444 and as required for electrical characteristics. Material as required for plenum flame test and for electrical characteristics.

Color Code: Per TIA 568 for Category 3 cable, and manufacturer's specification.

Assembly: Insulated conductors twisted into pairs. Multiple pairs cabled together. Unshielded Ripcord under the jacket to facilitate removal.

Jacket: Material suitable for jacketing and as required to pass NFPA 262 for plenum rating. Thickness according to UL 444. Color: White unless specified.

Diameter: Nominal diameter shall not exceed that in table 333UP Column B.

Flame test: Passes NFPA 262 (formerly UL 910) for use in plenum spaces.

Electricals: EIA-568 Category 3.

Voltage: 300V but not printed on the jacket per UL 444.

Application: Communication Cable per NEC Art. 800 and UL 444. EIA-568 Category 3.

Installation: Suitable for open cabling use in plenum spaces as CMP, CL3P, or CL2P; also for open cabling in risers and horizontal runs since rated higher than those requirements.

Temp.: 0C to 60C minimum range, continuous operating ambient temperature.

QA-Flame: UL Listed as cable type CMP. Cable ratings shall be printed on the jacket. UL file number will be printed on the jacket alongside the gauge size and cable rating. UL label shall be displayed on every cable package near manufacturer's label.

QA-Elec.: NRTL certified Cat 3 (preferred.) Manufacturer self-certified Cat 3 accepted.

Packaging: 1000ft or 500ft: spools, reels or boxes.

Appr. Manf: Manufacturer make and model shall be as listed in table 333UP.

Availability: Cable shall be available from local distribution known to project management.

333UP: 300V Specification 333: Paired Solid BC **Category 3**

Const	SDN	B	C	TAPPAN	GENERAL CABLE	BELDEN-WEST PENN	BELDEN
24/2pr	333UP.2	.156	12	I19991.1	2131243	-----	1243A2
24/3pr	333UP.3	.167	16	I19987.1	2131244	-----	24571097
24/4pr	333UP.4	.182	21	I19989.1	2131245	-----	1245A2
24/6pr	333UP.6	.211	28	I19992.1	2131246	-----	------
24/12pr	333UP,12	.275	52	I19994.1	-----	-----	------
24/25pr	333UP.25	.420	102	I19985.1	2131505	WP56801	24571221
24/50pr	333UP.50	.590	224	I90004.1	2131757	WP56126	24571235
24/100pr	333UP.100	.840	450	I90005.1	2131758	WP56128	24571250
24/200pr	333UP.200	1.100	852	I90006.1	2131442	WP56129	24571265
24/300pr	333UP.300	1.450	1243	-----	2131474	-----	24571266

Manufacturer's makes vary. Diameters (Col. B) and weights (Col. C) are conservative and for engineering purposes only. Consult manufacturer's data for individual make & model information.

Column B: Nominal Cable Diameter in inches, for conduit fill calculations.
Column C: Nominal cable weight in pounds per 1000ft.

335UH: 300V Specification 335: Paired Solid BC Category 5e
Solid bare copper / 100Ω Multiple Pairs / Unshielded (U) / Category 5e Horizontal Rated (R).

Conductor: 24 AWG Solid Bare Copper.

Insulation: Thickness according to UL 444 and as required for electrical characteristics. Material as required for flame test and for electrical characteristics.

Color Code: Pr 1 Orange & White w/Orange Stripe; Pr 2 Green & White w/Green Stripe; Pr 3 Blue & White w/Blue Stripe; Pr 4 Brown & White w/Brown Stripe. Per EIA-568 for Category 5e, or according to manufacturer's specification.

Assembly: Insulated conductors twisted into pairs. Multiple pairs cabled together. Unshielded. Ripcord under the jacket to facilitate removal.

Jacket: PVC. Thickness according to UL 444. Available in multiple colors.

Diameter: Nominal diameter shall not exceed that in table 335UH Column B.

Flame test: Passes UL 1685 Vertical Tray (UL 1581) for CM rated models. Passes UL 1666 for CMR rated models.

Electricals: EIA-568 Category 5e.

Voltage: 300V but not printed on the jacket per UL 444.

Application: Communication Cable per NEC Art. 800 and UL 444. Category 5e per EIA-568.

Installation: Models rated CMR are suitable for use in risers as open cabling. Models rated CM are suitable for use in horizontal runs as open cabling. Not for use in plenum spaces unless running in conduit.

Temp.: 0C to 60C minimum range, continuous operating ambient temperature.

QA-Flame: UL Listed as cable type CM or CMR. Cable ratings shall be printed on the jacket. UL file number will be printed on the jacket alongside the gauge size and cable rating. UL label shall be displayed on every cable package near manufacturer's label.

QA-Elec: NRTL certified Cat 5e (preferred.) Manufacturer self-certified Cat 5e accepted.

Packaging: 1000ft or 500ft: spools, reels or boxes.

Appr. Manf: Manufacturer make and model shall be as listed in table 335UH.

Availability: Cable shall be available from local distribution known to project management.

335UH: 300V Specification 335: Paired Solid BC Category 5e

Const	SDN	B	C	TAPPAN	GENERAL CABLE	BELDEN-WEST PENN	BELDEN
24/2pr	335UH.2	.157	12	I19977	-----	-----	1588A
24/3pr	335UH.3	.168	16	I19976	-----	------	------
24/4pr	335UH.4	.191	21	R19776	5133200e 6133714	4245	1583A 1500A
24/8pr	335UH.8	.2 x.4	42	H91868	-----	-----	-----
24/25pr	335UH.25	.420	105	-----	2133269e	WP58141	IBDN25R

Manufacturer's makes vary. Diameters (Col. B) and weights (Col. C) are conservative and for engineering purposes only. Consult manufacturer's data for individual make & model information.

Column B: Nominal Cable Diameter in inches, for conduit fill calculations.
Column C: Nominal cable weight in pounds per 1000ft.

335UP: 300V Specification 335: Paired Solid BC Category 5e
Solid bare copper / 100Ω Multiple Pairs / Unshielded (U) / Category 5e Plenum Rated (P).

Conductor: 24 AWG Solid Bare Copper.

Insulation: Thickness according to UL 444 and as required for electrical characteristics. Material as required for plenum flame test and for electrical characteristics.

Color Code: Pr 1 Orange & White w/Orange Stripe; Pr 2 Green & White w/Green Stripe; Pr 3 Blue & White w/Blue Stripe; Pr 4 Brown & White w/Brown Stripe. Per EIA-568 for Category 5e, or according to manufacturer's specification.

Assembly: Insulated conductors twisted into pairs. Multiple pairs cabled together. Unshielded. Ripcord under the jacket to facilitate removal.

Jacket: LS-PVC. Thickness according to UL 444. Available in multiple colors.

Diameter: Nominal diameter shall not exceed that in table 335UP Column B.

Flame test: Passes NFPA 262 (formerly UL 910) for use in plenum spaces.

Electricals: EIA-568 Category 5e.

Voltage: 300V but not printed on the jacket per UL 444.

Application: Communication Cable per NEC Art. 800 and UL 444. Category 5e per EIA-568.

Installation: Suitable for open cabling use in plenum spaces as CMP, CL3P, or CL2P; also for open cabling in risers and horizontal runs since rated higher than those requirements.

Temp.: 0C to 60C minimum range, continuous operating ambient temperature.

QA-Flame: UL Listed as cable type CMP. Cable ratings shall be printed on the jacket. UL file number will be printed on the jacket alongside the gauge size and cable rating. UL label shall be displayed on every cable package near manufacturer's label.

QA-Elec: NRTL certified Cat 5e (preferred.) Manufacturer self-certified Cat 5e accepted.

Packaging: 1000ft or 500ft: spools, reels or boxes.

Appr. Manf: Manufacturer make and model shall be as listed in table 335UP.

Availability: Cable shall be available from local distribution known to project management.

335UP: 300V Specification 335: Paired Solid BC Category 5e

Const	SDN	B	C	TAPPAN	GENERAL CABLE	BELDEN-WEST PENN	BELDEN
24/2pr	335UP.2	.156	12	I19975	-----	-----	1590A
24/3pr	335UP.3	.167	16	I19974	-----	-----	
24/4pr	335UP.4	.182	21	P19915	5131361e 6133716	25425	1585A 1501A
24/25pr	335UP.25	.420	102	-----	2133694e	WP58142	IBDN25P

Manufacturer's makes vary. Diameters (Col. B) and weights (Col. C) are conservative and for engineering purposes only. Consult manufacturer's data for individual make & model information.

Column B: Nominal Cable Diameter in inches, for conduit fill calculations.
Column C: Nominal cable weight in pounds per 1000ft.

336UH: 300V Specification 336: Paired Solid BC Category 6
Solid bare copper / 100Ω Multiple Pairs / Unshielded (U) / Category 6 Horizontal Rated (H).

Conductor: 23 AWG Solid Bare Copper.

Insulation: Thickness according to UL 444 and as required for electrical characteristics. Material as required for flame test and for electrical characteristics.

Color Code: Pr 1 Orange & White w/Orange Stripe; Pr 2 Green & White w/Green Stripe; Pr 3 Blue & White w/Blue Stripe; Pr 4 Brown & White w/Brown Stripe. Per EIA-568 for Category 6, or according to manufacturer's specification.

Assembly: Insulated conductors twisted into pairs. Multiple pairs cabled together. Unshielded. A spline or other means of adding physical pair spacing is optional, SDN 336UH.4B Ripcord under the jacket to facilitate removal.

Jacket: PVC. Thickness according to UL 444. Available in multiple colors.

Diameter: Nominal diameter shall not exceed that in table 336UH Column B.

Flame test: Passes UL 1685 Vertical Tray (UL 1581) for CM rated models. Passes UL 1666 for CMR rated models.

Electricals: EIA-568 Category 6.

Voltage: 300V but not printed on the jacket per UL 444.

Application: Communication Cable per NEC Art. 800 and UL 444. Category 6 per EIA-568.

Installation: Models rated CMR are suitable for use in risers as open cabling. Models rated CM are suitable for use in horizontal runs as open cabling. Not for use in plenum spaces unless running in conduit.

Temp.: 0C to 60C minimum range, continuous operating ambient temperature.

QA-Flame: UL Listed as cable type CM or CMR. Cable ratings shall be printed on the jacket. UL file number will be printed on the jacket alongside the gauge size and cable rating. UL label shall be displayed on every cable package near manufacturer's label.

QA-Elec: NRTL certified Cat 6 (preferred.) Manufacturer self-certified Cat 6 accepted.

Packaging: 1000ft or 500ft: spools, reels or boxes.

Appr. Manf: Manufacturer make and model shall be as listed in table 336UH.

Availability: Cable shall be available from local distribution known to project management.

336UH: 300V Specification 336: Paired Solid BC Category 6

Const	SDN	Compliant Frequency	B	C	TAPPAN	GENERAL CABLE	BELDEN-WEST PENN	BELDEN
336UH/4pr	336UH.4A	250MHz	.230	26	I99995	7133767	4246	7881A
336UH/4pr	336UH.4B	≥500MHz*	.240	32	I99998	7133329 7133724	-----	24566345 24586985

Manufacturer's makes vary. Diameters (Col. B) and weights (Col. C) are conservative and for engineering purposes only. Consult manufacturer's data for individual make & model information.

Column B: Nominal Cable Diameter in inches, for conduit fill calculations.
Column C: Nominal cable weight in pounds per 1000ft.

* Compliance to higher frequency for some, but not necessarily all, significant EIA-568 Category 6 electrical characteristics.

336UP: 300V Specification 336: Paired Solid BC Category 6
Solid bare copper / 100Ω Multiple Pairs / Unshielded (U) / Category 6 Plenum Rated (P).

Conductor: 23 AWG Solid Bare Copper.

Insulation: Thickness according to UL 444 and as required for electrical characteristics. Material as required for plenum flame test and for electrical characteristics.

Color Code: Pr 1 Orange & White w/Orange Stripe; Pr 2 Green & White w/Green Stripe; Pr 3 Blue & White w/Blue Stripe; Pr 4 Brown & White w/Brown Stripe. Per EIA-568 for Category 5e, or according to manufacturer's specification.

Assembly: Insulated conductors twisted into pairs. Multiple pairs cabled together. Unshielded. A spline or other means of adding physical pair spacing is optional, SDN 336UP.4B Ripcord under the jacket to facilitate removal.

Jacket: LS-PVC. Thickness according to UL 444. Available in multiple colors.

Diameter: Nominal diameter shall not exceed that in table 336UP Column B.

Flame test: Passes NFPA 262 (formerly UL 910) for use in plenum spaces.

Electricals: EIA-568 Category 6.

Voltage: 300V but not printed on the jacket per UL 444.

Application: Communication Cable per NEC Art. 800 and UL 444. Category 6 per EIA-568.

Installation: Suitable for open cabling use in plenum spaces as CMP, CL3P, or CL2P; also for open cabling in risers and horizontal runs since rated higher than those requirements.

Temp.: 0C to 60C minimum range, continuous operating ambient temperature.

QA-Flame: UL Listed as cable type CMP. Cable ratings shall be printed on the jacket. UL file number will be printed on the jacket alongside the gauge size and cable rating. UL label shall be displayed on every cable package near manufacturer's label.

QA-Elec: NRTL certified Cat 6 (preferred.) Manufacturer self-certified Cat 6 accepted.

Packaging: 1000ft or 500ft: spools, reels or boxes.

Appr. Manf: Manufacturer make and model shall be as listed in table 336UP.

Availability: Cable shall be available from local distribution known to project management.

336UP: 300V Specification 336: Paired Solid BC Category 6

Const	SDN	Compliant Frequency	B	C	TAPPAN	GENERAL CABLE	BELDEN-WEST PENN	BELDEN
336UP/4pr	336UP.4A	250MHz	.230	26	I99993	7131761	254246	7882A
336UP/4pr	335UP.4B	≥500MHz*	.240	32	I99997	7131450 7131722	-----	24567345 24587385

Manufacturer's makes vary. Diameters (Col. B) and weights (Col. C) are conservative and for engineering purposes only. Consult manufacturer's data for individual make & model information.

Column B: Nominal Cable Diameter in inches, for conduit fill calculations.
Column C: Nominal cable weight in pounds per 1000ft.

* Compliance to higher frequency for some, but not necessarily all, significant EIA-568 Category 6 electrical characteristics.

341UH: 300V Specification 341: Solid Bare Copper Fire Alarm Cables
Solid BC / Cabled / Fire Alarm / Unshielded (U) / Jacket

Unshielded
Horizontal Rated (H).
or Riser Rated (R).

Conductor: Solid Bare Copper.

Insulation: Material per Table 341UH. Thickness according to UL 1424 or as needed to achieve the desired capacitance.

Color Code: Black, Red, White, Green, Brown, Blue, Orange, Yellow, Violet, Gray, Pink, Beige.

Assembly: Twisted together. 2C is 1 pair, 3C is 1 Triad, and 4C is one quad.
Ripcord under the jacket to facilitate removal.

Shielding: None.

Jacket: PVC. Thickness according to UL 1424.
Color: Red unless specified.

Diameter: Nominal diameter shall not exceed that in table 341UH Column B.

Flame test: Passes UL 1685 Vertical Tray (UL 1581) for FPL rated models.
Passes UL 1666 for FPLR rated models.

Voltage: 300V but not printed on the jacket per UL 1424.

Application: Fire Power Limited Circuit Cable per NEC Art. 760 and UL 1424.

Mid-Cap FA: Data Grade: Addressable: Cables with capacitance 35pF/ft or lower are suitable for use with Addressable Fire Alarm systems in installations calling for "Mid-Cap" Fire Alarm Cable or "Data Grade" Fire Alarm cable. Consult device manufacturers' instructions and "CAP" column in table 341UH. .

Installation: Models rated FPLR are suitable for use in risers and horizontal runs as open cabling.
Models rated FPL are suitable for use only in horizontal runs as open cabling.
Not for use in plenum spaces unless running in conduit.

Temp.: 0C to 60C minimum range, continuous operating ambient temperature.

QA: UL Listed as cable type FPL or FPLR. Cable ratings shall be printed on the jacket.
UL file number will be printed on the jacket alongside the gauge size and cable rating.
UL label shall be displayed on every cable package near manufacturer's label.

Packaging: 1000ft or 500ft: spools, reels or boxes.

Appr. Manf: Manufacturer make and model shall be as listed in table 341UH.

Availability: Cable shall be available from local distribution known to project management

Table 341UH - Solid BC, Cabled Unshielded Fire Alarm Cable, Horizontal or Riser Rated

Const	SDN	A	INS	JKT	CAP	TEMP	B	C	TAPPAN	GENERAL CABLE	BELDEN West Penn	BELDEN
22/2	341UR.222C	1	3	1	2	75C	.121	9	F20005	E1482S	-----	-----
22/4	341UR.224C	1	3	1	2	75C	.140	16	F20004	E1484S	-----	5522UL
22/6	341UR.226C	1	3	1	2	75C	.168	21	F20009	E1486S	-----	5524UL
22/8	341UR.228C	1	3	1	2	75C	.170	25	F20002	----	-----	-----
18/2	341UH.182A	1	1	1	1	75C	.206	22	-----	-----	-----	5320UJ
	341UH.182B	1	2	1	2	75C	.182	19	SM4007	C0471	D980	-----
	341UR.182C	1	3	1	3	75C	.151	17	F40021	E1502S	980	5320UL
18/4	341UH.184B	1	2	1	2	75C	.230	38	SM4009	C0485	D982	-----
	341UR.184C	1	3	1	3	75C	.176	30	F40568	E1504S	982	5322UL
18/6	341UR.186C	1	3	1	3	75C	.212	44	F40025	E1506S	-----	5324UL
18/8	341UR.188C	1	3	1	3	75C	.230	61	F40026	E1508S	-----	5326UL
18/10	314UR.1810C	1	3	1	3	75C	.272	71	F40027	-----	-----	5328UL
18/12	314UR.1812C	1	3	1	3	75C	.281	83	-----	-----	-----	5329UL
16/2	341UH.162A	1	1	1	1	75C	.230	32	-----	-----	-----	5220UJ
	341UH.162B	1	2	1	2	75C	.202	27	SM5014	C0473	D990	-----
	341UR.162C	1	3	1	3	75C	.180	25	F50004	E1512S	990	5220UL
16/4	341UR.164C	1	3	1	3	75C	.204	45	F50383	E1514S	992	5222UL
14/2	341UH.142B	1	2	1	2	75C	.240	42	SM6005	C0491	-----	-----
	341UR.142C	1	3	1	3	75C	.213	38	F60001	E1522S	994	5120UL
14/4	341UR.144C	1	3	1	3	75C	.251	70	F60355	E1524S	700	5122UL
12/2	341UH.122B	1	2	1	2	75C	.282	60	SM7001	C0492	----	----
	341UR.122C	1	3	1	3-4	75C	.248	55	F70004	E1532S	998	5020UL
12/4	341UR.124C	1	3	1	3	75C	.293	101	F70006	E1534S	----	----

Manufacturer's makes vary. Diameters (Col. B) and weights (Col. C) are conservative and for engineering purposes only. Consult manufacturer's data for individual make & model information.

Column A:	Solid copper. For stranded Fire Alarm Cable see Specification 42.
INS:	'1'=Foam Polyethylene (FPE); '2'=Polyethylene, '3'=PVC or Polypropylene
JKT:	'1'=PVC
CAP:	Capacitance in pF/ft '1' = <16 '2'= 16-24 '3'=25-35 '4'=36-49 '5'= 50 & up
Column B:	Cable Diameter in inches. Manufacturers' vary. Consult manufacturer's spec.'s.
Column C:	Nominal cable weight in pounds per 1000ft

<u>Mid-Cap FA</u>: <u>Data Grade:</u> <u>Addressable:</u> Cables with capacitance 35pF/ft or lower are suitable for use with Addressable Fire Alarm systems in installations calling for "Mid-Cap" Fire Alarm Cable or "Data Grade" Fire Alarm cable. Consult device manufacturers' instructions and "CAP" column in table 341UH.

341UP: 300V Specification 341: Solid Bare Copper Fire Alarm Cables **Unshielded**
Solid BC / Cabled / Fire Alarm / Unshielded (U) / Jacket **Plenum Rated (P).**

Conductor:	Solid Bare Copper.
Insulation:	Material per Table 341UP. Thickness according to UL 1424 or as needed to achieve the desired capacitance.
Color Code:	Black, Red, White, Green, Brown, Blue, Orange, Yellow, Violet, Gray, Pink, Beige.
Assembly:	Twisted together. 2C is 1 pair, 3C is 1 Triad, and 4C is one quad. Ripcord under the jacket to facilitate removal.
Shielding:	None.
Jacket:	LS-PVC. Thickness according to UL 1424. Color: Red unless specified.
Diameter:	Nominal diameter shall not exceed that in table 341UP Column B.
Flame test:	Passes NFPA 262 (formerly UL 910).
Voltage:	300V but not printed on the jacket per UL 1424.
Application:	Fire Power Limited Circuit Cable per NEC Art. 760 and UL 1424.
Mid-Cap FA: Data Grade: Addressable:	Cables with capacitance 35pF/ft or lower are suitable for use with Addressable Fire Alarm systems in installations calling for “Mid-Cap” Fire Alarm Cable or “Data Grade” Fire Alarm cable. Consult device manufacturers’ instructions and “CAP” column in table 341UP.
Installation:	Suitable for use as open cabling in plenum spaces, type FPLP. Also for use in risers and in horizontal runs since rated higher than those requirements.
Temp.:	0C to 60C minimum range, continuous operating ambient temperature.
QA:	UL Listed as cable type FPLP. Cable ratings shall be printed on the jacket. UL file number will be printed on the jacket alongside the gauge size and cable rating. UL label shall be displayed on every cable package near manufacturer’s label.
Packaging:	1000ft or 500ft: spools, reels or boxes.
Appr. Manf:	Manufacturer make and model shall be as listed in table 341UP.
Availability:	Cable shall be available from local distribution known to project management.

Table 341UP - Solid BC, Cabled Unshielded Fire Alarm Cable, Plenum Rated

Const	SDN	A	INS	JKT	CAP	TEMP	B	C	TAPPAN	GENERAL CABLE	BELDEN West Penn	BELDEN
22/2	341UP.222	1	3	3	3	75C	.121	12	G20002	E3482S	-----	-----
22/4	341UP.224	1	3	3	3	75C	.137	16	G20342	E3484S	-----	6522UL
22/6	341UP.226	1	3	3	3	75C	.162	22	G20006	-----	-----	6524UL
22/8	341UP.228	1	3	3	3	75C	.166	26	G20008	-----	-----	-----
22/10	341UP.2210	1	3	3	3	75C	.178	31	G20010	-----	-----	-----
22/12	341UP.2212	1	3	3	3	75C	.192	37	G20012	-----	-----	-----
18/2	341UP.182A	1	2	3	2-3	75C	.172	20	J40028	C3240	-----	6320UJ
	341UP.182B	1	3	3	3	75C	.155	18	G40005	E3502S	60980B	6320UL
18/3	341UP.183B	1	3	3	3	75C	.160	28	G40039	E3503S	-----	6321UL
18/4	341UP.184A	1	2	3	2	75C	.195	33	J40030	C3242	-----	-----
	341UP.184B	1	3	3	3	75C	.183	31	G40505	E3504S	60982B	6322UL
18/6	341UP.186B	1	3	3	3	75C	.211	46	G40503	E3506S	-----	6324UL
18/8	341UP.188B	1	3	3	3	75C	.230	63	G40501	-----	-----	6326UL
18/10	341UP.1810B	1	3	3	3	75C	.272	73	G40045	-----	-----	6328UL
16/2	341UP.162A	1	2	3	2-3	75C	.196	27	J50009	C3241	-----	6220UJ
	341UP.162B	1	3	3	3-4	75C	.174	26	G50032	E3512S	60991B	6220UL
16/4	341UP.164A	1	2	3	2	75C	.231	57	-----	C3243	-----	-----
	341UP.164B	1	3	3	3-4	75C	.204	47	G50370	E3514S	60164B	6222UL
14/2	341UP.142A	1	2	3	2-3	75C	.242	40	J60004	C3244	-----	6120UJ
	341UP.142B	1	3	3	3-4	75C	.205	38	G60004	E3522S	60993B	6120UL
14/4	341UP.144A	1	2	3	2	75C	.286	80	-----	C3245	-----	-----
	341UP.144B	1	3	3	3-4	75C	.241	70	G60347	E3524S	60700B	6122UL
12/2	341UP.122A	1	2	3	2	75C	.282	60	J70004	C3246	-----	-----
	341UP.122B	1	3	3	4	75C	.246	59	G70009	E3532S	60995B	6020UL
12/4	341UP.124A	1	2	3	2	75C	.333	115	-----	C3247	-----	-----
	341UP.124B	1	3	3	4	75C	.291	104	G70127	E3534S	-----	-----

Manufacturer's makes vary. Diameters (Col. B) and weights (Col. C) are conservative and for engineering purposes only. Consult manufacturer's data for individual make & model information.

INS: '1'=Foam FEP or Foam Fluoropolymer, '2'=FEP or other fluoropolymer, '3'=LS-PVC
JKT: '1'=FEP. '2'=other fluoropolymer. '3'= LS-PVC
CAP: Capacitance in pF/ft '1' = <16 '2'= 16-24 '3"=25-35 '4'=36-49 '5'= 50 & up
Column B: Cable Diameter in inches. Manufacturers' vary. Consult manufacturer's spec.'s.
Column C: Nominal cable weight in pounds per 1000ft

Mid-Cap FA: Data Grade: Addressable: Cables with capacitance 35pF/ft or lower are suitable for use with Addressable Fire Alarm systems in installations calling for "Mid-Cap" Fire Alarm Cable or "Data Grade" Fire Alarm cable. Consult device manufacturers' instructions and "CAP" column in table 341UP.

341SH: 300V Specification 341: Solid Bare Copper Fire Alarm Cables
Solid BC / Cabled / Fire Alarm / Shielded (S) / Jacket

Shielded
Horizontal Rated (H).
or Riser Rated (R).

Conductor: Solid Bare Copper.

Insulation: Material per Table 341SH. Thickness according to UL 1424 or as needed to achieve the desired capacitance.

Color Code: Black, Red, White, Green, Brown, Blue, Orange, Yellow, Violet, Gray, Pink, Beige.

Assembly: Twisted together. 2C is 1 pair, 3C is 1 Triad, and 4C is one quad.
Ripcord under the jacket to facilitate removal.

Shielding: Shielded overall with aluminum-polyester tape and a drain wire.

Jacket: PVC. Thickness according to UL 1424.
Color: Red unless specified.

Flame test: Passes UL 1685 Vertical Tray (UL 1581) for FPL rated models.
Passes UL 1666 for FPLR rated models.

Voltage: 300V but not printed on the jacket per UL 1424.

Application: Fire Power Limited Circuit Cable per NEC Art. 760 and UL 1424.

Mid-Cap FA: Data Grade: Addressable: Cables with capacitance 35pF/ft or lower are suitable for use with Addressable Fire Alarm systems in installations calling for “Mid-Cap” Fire Alarm Cable or “Data Grade” Fire Alarm cable. Consult device manufacturers’ instructions and “CAP” column in table 341SH.

Installation: Models rated FPLR are suitable for use in risers and horizontal runs as open cabling.
Models rated FPL are suitable for use only in horizontal runs as open cabling.
Not for use in plenum spaces unless running in conduit.

Temp.: 0C to 60C minimum range, continuous operating ambient temperature.

QA: UL Listed as cable type FPL or FPLR. Cable ratings shall be printed on the jacket.
UL file number will be printed on the jacket alongside the gauge size and cable rating.
UL label shall be displayed on every cable package near manufacturer’s label.

Packaging: 1000ft or 500ft: spools, reels or boxes.

Appr. Manf: Manufacturer make and model shall be as listed in table 341SH.

Availability: Cable shall be available from local distribution known to project management

Table 341SH - Solid BC, Cabled Shielded Fire Alarm Cable, Horizontal or Riser Rated
SDN with suffix A-C have lower capacitance for addressable systems / data transmission
SDN with suffix D are rated FPLR for use in Risers.

Const	SDN	A	INS	JKT	CAP	TEMP	B	C	TAPPAN	GENERAL CABLE	BELDEN West Penn	BELDEN
22/2	341SR.222D	1	3	1	4	75C	.126	12	F20015	E2482S	-----	-----
22/4	341SR.224D	1	3	1	4	75C	.145	18	F20017	E2484S	-----	5522FL
18/2	341SH.182A	1	1	1	3	75C	.211	26	-----	-----	-----	5320FJ
	341SH.182B	1	2	1	3	75C	.181	21	J40026	-----	-----	-----
	341SH.182C	1	3	1	3	60C	.210	27	-----	C0472	D975	-----
	341SR.182D	1	3	1	5	75C	.158	23	F40015	E2502S	975	5320FL
18/4	341SH.184A	1	1	1	2	75C	.240	30	-----	-----	-----	5322FJ
	341SH.184B	1	2	1	3	75C	.230	42	SM4012	-----	-----	
	341SH.184C	1	3	1	3	60C	.210	36	-----	C0494	-----	-----
	341SR.184D	1	3	1	5	75C	.183	35	F40567	E2504S	977	5322FL
18/6	341SR.186D	1	3	1	5	75C	.216	49	F40032	E2506S	-----	-----
18/8	341SR.188D	1	3	1	5	75C	.235	60	F40033	E2508S	-----	-----
16/2	341SH.162A	1	1	1	3	75C	.235	29	-----	-----	-----	5220FJ
	341SH.162B	1	2	1	3	75C	.202	29	SM5015	-----	-----	-----
	341SH.162C	1	3	1	3	60C	.226	33	-----	C0474	D991	-----
	341SR.162D	1	3	1	5	75C	.180	32	F50010	E2522S	991	5220FL
16/4	341SH.164A	1	1	1	2	75C	.269	50	-----	-----	-----	5222FJ
	341SH.164B	1	2	1	3	75C	.256	56	SM5019		-----	-----
	341SH.164C	1	3	1	3	60C	.246	52	-----	C0495	-----	-----
	341SR.164D	1	3	1	5	75C	.210	48	F50382	E2524S	993	5222FL
14/2	341SH.142A	1	1	1	3	75C	.279	45	-----		-----	5120FJ
	341SH.142B	1	2	1	3	75C	.248	43	SM6008	-----	-----	-----
	341SH.142C	1	3	1	3	60C	.262	47	-----	C0475	D995	-----
	341SR.142D	1	3	1	5	75C	.218	43	F60030	E2532S	995	5120FL
14/4	341SH.144C	1	3	1	3	60C	.287	76	-----	C0496	-----	-----
	341SR.144D	1	3	1	5	75C	.255	72	F60354	E2534S	-----	5122FL
12/2	341SH.122A	1	1	1	3	75C	.317	64	-----	----	----	5020FJ
	341SH.122B	1	2	1	3-4	75C	.262	58	SM7003		----	-----
	341SH.122C	1	3	1	3	60C	.287	63	-----	C0476	----	----
	341SR.122D	1	3	1	5	75C	.252	59	F70015	E2542S	999	5020FL
12/4	341SH.124C	1	3	1	3	60C	.337	112	-----	C0497	-----	-----
	341SR.124D	1	3	1	5	75C	.303	108	F70137	E2544S	-----	-----

Manufacturer's makes vary. Diameters (Col. B) and weights (Col. C) are conservative and for engineering purposes only. Consult manufacturer's data for individual make & model information.

Column A: Minimum number of strands. Consult manufacturer's spec.'s for higher strands.
INS: '1'=Foam Polyethylene (FPE); '2'=Polyethylene, '3'=PVC or Polypropylene
JKT: '1'=PVC
CAP: Capacitance in pF/ft '1' = <16 '2'= 16-24 3"=25-35 '4'=36-49 '5'= 50 & up
Column B: Cable Diameter in inches. Manufacturers' vary. Consult manufacturer's spec.'s.
Column C: Nominal cable weight in pounds per 1000ft

Mid-Cap FA: Data Grade: Addressable: Cables with capacitance 35pF/ft or lower are suitable for use with Addressable Fire Alarm systems in installations calling for "Mid-Cap" Fire Alarm Cable or "Data Grade" Fire Alarm cable. Consult device manufacturers' instructions and "CAP" column in table 341SH.

341SP: 300V Specification 341: Solid Bare Copper Fire Alarm Cables **Shielded**
Solid BC / Cabled / Fire Alarm / Shielded (S) / Jacket. **Plenum Rated (P).**

Conductor:	Solid Bare Copper.
Insulation:	Material per Table 341SP. Thickness according to UL 1424 or as needed to achieve the desired capacitance.
Color Code:	Black, Red, White, Green, Brown, Blue, Orange, Yellow, Violet, Gray, Pink, Beige.
Assembly:	Twisted together. 2C is 1 pair, 3C is 1 Triad, and 4C is one quad. Ripcord under the jacket to facilitate removal.
Shielding:	Shielded overall with aluminum-polyester tape and a drain wire.
Jacket:	LS-PVC. Thickness according to UL 1424. Color: Red unless specified.
Diameter:	Nominal diameter shall not exceed that in table 341SP Column B.
Flame test:	Passes NFPA 262 (formerly UL 910).
Voltage:	300V but not printed on the jacket per UL 1424.
Application:	Fire Power Limited Circuit Cable per NEC Art. 760 and UL 1424.
Mid-Cap FA: Data Grade: Addressable:	Cables with capacitance 35pF/ft or lower are suitable for use with Addressable Fire Alarm systems in installations calling for “Mid-Cap” Fire Alarm Cable or “Data Grade” Fire Alarm cable. Consult device manufacturers’ instructions and “CAP” column in table 341SP.
Installation:	Suitable for use as open cabling in plenum spaces, type FPLP. Also for use in risers and in horizontal runs since rated higher than those requirements.
Temp.:	0C to 60C minimum range, continuous operating ambient temperature.
QA:	UL Listed as cable type FPLP. Cable ratings shall be printed on the jacket. UL file number will be printed on the jacket alongside the gauge size and cable rating. UL label shall be displayed on every cable package near manufacturer’s label.
Packaging:	1000ft or 500ft: spools, reels or boxes.
Appr. Manf:	Manufacturer make and model shall be as listed in table 341SP.
Availability:	Cable shall be available from local distribution known to project management.

Table 341SP - Solid BC; Cabled; Shielded; Fire Alarm Cable; Plenum Rated

Const	SDN	A	INS	JKT	CAP	TEMP	B	C	TAPPAN	GENERAL CABLE	BELDEN West Penn	BELDEN
22/2	341SP.222	1	3	3	4	75C	.126	13	G20013	E3542S	-----	-----
22/4	341SP.224	1	3	3	4	75C	.128	16	G20338	-----	-----	-----
22/6	341SP.226	1	3	3	4	75C	.144	22	G20017	-----	-----	-----
18/2	341SP.182A	1	2	3	3	75C	.185	23	J40002	C3167	D60975	6320FK
	341SP.182B	1	3	3	5	75C	.158	22	G40006	E3602S	60975B	6320FL
18/3	341SP.183B	1	3	3	5	75C	.165	28	G40048	E3603S	-----	-----
18/4	341SP.184A	1	2	3	3	75C	.202	36	-----	C3170	-----	-----
	341SP.184B	1	3	3	5	75C	.183	33	G40504	E3604S	60977B	6322FL
18/6	341SP.186B	1	3	3	5	75C	.216	48	G40502	E6306S	-----	-----
16/2	341SP.162A	1	2	3	3-4	75C	.210	28	J50010	C3168	D60990	6220FK
	341SP.162B	1	3	3	5	75C	.180	30	G50013	E3612S	60990B	6220FL
16/4	341SP.164A	1	2	3	3	75C	.238	52	-----	C3171	-----	-----
	341SP.164B	1	3	3	5	75C	.210	50	G50369	E3614S	603164B	6222FL
14/2	341SP.142A	1	2	3	3-4	75C	.247	44	J60005	C3172	-----	-----
	341SP.142B	1	3	3	5	75C	.210	41	G60025	E3622S	60992B	6120FL
14/4	341SP.144A	1	2	3	3	75C	.289	80	-----	C3173	-----	-----
	341SP.144B	1	3	3	4	75C	.248	74	G60346	E3624S	-----	6122FL
12/2	341SP.122A	1	2	3	4	75C	.289	66	-----	C3174	-----	-----
	341SP.122B	1	3	3	5	75C	.252	61	G70019	E3632S	60994B	6020FL
12/4	341SP.124A	1	2	3	3	75C	.340	118	-----	C3175	-----	-----
	341SP.124B	1	3	3	5	75C	.300	112	G70021	E3634S	-----	-----

Manufacturer's makes vary. Diameters (Col. B) and weights (Col. C) are conservative and for engineering purposes only. Consult manufacturer's data for individual make & model information.

INS:	'1'=Foam FEP or Foam Fluoropolymer, '2'=FEP or other fluoropolymer, '3'=LS-PVC
JKT:	'1'=FEP. '2'=other fluoropolymer. '3'= LS-PVC
CAP:	Capacitance in pF/ft '1' = <16 '2'= 16-24 '3''=25-35 '4'=36-49 '5'= 50 & up
Column B:	Cable Diameter in inches. Manufacturers' vary. Consult manufacturer's spec.'s.
Column C:	Nominal cable weight in pounds per 1000ft

Mid-Cap FA: Data Grade: Addressable:	Cables with capacitance 35pF/ft or lower are suitable for use with Addressable Fire Alarm systems in installations calling for "Mid-Cap" Fire Alarm Cable or "Data Grade" Fire Alarm cable. Consult device manufacturers' instructions and "CAP" column in table 341SP.

342UR: 300V Specification 342: Stranded Bare Copper Fire Alarm Cables **Unshielded**
Stranded BC / Cabled / Fire Alarm / Unshielded (U) / Jacket **Riser Rated (R).**

Conductor: Stranded Bare Copper. Number of strands per Table 342UR Column A.

Insulation: PVC. Thickness according to UL 1424.

Color Code: Black, Red, White, Green.

Assembly: Twisted together. 2C is 1 pair, 3C is 1 Triad, and 4C is one quad.
Ripcord under the jacket to facilitate removal.

Shielding: None.

Jacket: PVC. Thickness according to UL 1424.
Color: Red unless specified.

Diameter: Nominal diameter shall not exceed that in table 342UR Column B.

Flame test: Passes UL 1666.

Voltage: 300V but not printed on the jacket per UL 1424.

Application: Fire Power Limited Circuit Cable per NEC Art. 760 and UL 1424.

Installation: Suitable for use in risers and horizontal runs as open cabling. Type FPLR.

Temp.: 0C to 60C minimum range, continuous operating ambient temperature.

QA: UL Listed as cable type FPLR. Cable ratings shall be printed on the jacket.
UL file number will be printed on the jacket alongside the gauge size and cable rating.
UL label shall be displayed on every cable package near manufacturer's label.

Packaging: 1000ft or 500ft: spools, reels or boxes.

Appr. Manf: Manufacturer make and model shall be as listed in table 342UR.

Availability: Cable shall be available from local distribution known to project management

Table 342UR Str BC; Cabled; Fire Alarm; Unshielded (U); Riser (R) Rated.

Const	SDN	A	B	C	TAPPAN	BELDEN-WEST PENN	BELDEN
18/2	342UR.182	7	.158	17	F40142	-----	5300UL
18/4	342UR.184	7	.184	30	F40144	-----	5302UL
16/2	342UR.162	7	.182	26	F50289	990S	5200UL
16/4	342UR.164	7	.212	45	F50374	-----	5202UL
14/2	342UR.142	19	.224	38	F60340	994S	5100UL
14/4	342UR.144	19	.257	69	F60248	-----	5102UL
12/2	342UR.122	19	.266	56	F70092	998S	5000UL

Manufacturer's makes vary. Diameters (Col. B) and weights (Col. C) are conservative and for engineering purposes only. Consult manufacturer's data for individual make & model information.

Column A: Stranded bare copper conductor number of strands.
Column B: Nominal Cable Diameter in inches, for conduit fill calculations.
Column C: Nominal cable weight in pounds per 1000ft.

342UP: 300V Specification 342: Stranded Bare Copper Fire Alarm Cables Unshielded
Stranded BC / Cabled / Fire Alarm / Unshielded (U) / Jacket Plenum Rated (P).

Conductor: Stranded Bare Copper. Number of strands per Table 342UP Column A.

Insulation: LS-PVC. Thickness according to UL 1424.

Color Code: Black, Red, White, Green.

Assembly: Twisted together. 2C is 1 pair, 3C is 1 Triad, and 4C is one quad. Ripcord under the jacket to facilitate removal.

Shielding: None.

Jacket: LS-PVC. Thickness according to UL 1424. Color: Red unless specified.

Diameter: Nominal diameter shall not exceed that in table 342UP Column B.

Flame test: Passes NFPA 262 (formerly UL 910).

Voltage: 300V but not printed on the jacket per UL 1424.

Application: Fire Power Limited Circuit Cable per NEC Art. 760 and UL 1424.

Installation: Suitable for use as open cabling in plenum spaces, type FPLP. Also for use in risers and in horizontal runs since rated higher than those requirements.

Temp.: 0C to 60C minimum range, continuous operating ambient temperature.

QA: UL Listed as cable type FPLP. Cable ratings shall be printed on the jacket. UL file number will be printed on the jacket alongside the gauge size and cable rating. UL label shall be displayed on every cable package near manufacturer's label.

Packaging: 1000ft or 500ft: spools, reels or boxes.

Appr. Manf: Manufacturer make and model shall be as listed in table 342UP.

Availability: Cable shall be available from local distribution known to project management.

Table 342UP Str BC; Cabled; Fire Alarm; Unshielded (U); Plenum (P) Rated.

Const	SDN	A	B	C	TAPPAN	BELDEN-WEST PENN	BELDEN
18/2	342UP.182	7	.158	18	G40172	-----	6300UL
18/4	342UP.184	7	.185	31	G40174	-----	6302UL
16/2	342UP.162	7	.180	26	G50359	60991BS	6200UL
16/4	342UP.164	7	.210	47	G50160	-----	-----
14/2	342UP.142	19	.220	38	G60210	60993BS	6100UL
14/4	342UP.144	19	.255	71	G60120	-----	-----
12/2	342UP.122	19	.264	58	G70090	60995BS	-----

Manufacturer's makes vary. Diameters (Col. B) and weights (Col. C) are conservative and for engineering purposes only. Consult manufacturer's data for individual make & model information.

Column A: Stranded bare copper conductor number of strands.
Column B: Nominal Cable Diameter in inches, for conduit fill calculations.
Column C: Nominal cable weight in pounds per 1000ft.

342SR: 300V Specification 342: Stranded Bare Copper Fire Alarm Cables Shielded
Stranded BC / Cabled / Fire Alarm / Shielded (S) / Jacket Riser Rated (R).

Conductor: Stranded Bare Copper. Number of strands per Table 342SR Column A.

Insulation: PVC. Thickness according to UL 1424.

Color Code: Black, Red, White, Green.

Assembly: Twisted together. 2C is 1 pair, 3C is 1 Triad, and 4C is one quad.
Ripcord under the jacket to facilitate removal.

Shielding: Overall shield of aluminum-polyester foil and tinned copper drain wire.

Jacket: PVC. Thickness according to UL 1424.
Color: Red unless specified.

Diameter: Nominal diameter shall not exceed that in table 342SR Column B.

Flame test: Passes UL 1666.

Voltage: 300V but not printed on the jacket per UL 1424.

Application: Fire Power Limited Circuit Cable per NEC Art. 760 and UL 1424.

Installation: Suitable for use in risers and horizontal runs as open cabling. Type FPLR.

Temp.: 0C to 60C minimum range, continuous operating ambient temperature.

QA: UL Listed as cable type FPLR. Cable ratings shall be printed on the jacket.
UL file number will be printed on the jacket alongside the gauge size and cable rating.
UL label shall be displayed on every cable package near manufacturer's label.

Packaging: 1000ft or 500ft: spools, reels or boxes.

Appr. Manf: Manufacturer make and model shall be as listed in table 342SR.

Availability: Cable shall be available from local distribution known to project management

Table 342SR Str BC; Cabled; Fire Alarm; Shielded (S); Riser (R) Rated.

Const	SDN	A	B	C	TAPPAN	BELDEN-WEST PENN	BELDEN
18/2	342SR.182	7	.162	19	F40370	-----	5300FL
18/4	342SR.184	7	.188	30	-----	-----	5302FL
16/2	342SR.162	7	.184	27	F50375	-----	5200FL
16/4	342SR.164	7	.214	44	-----	-----	5202FL
14/2	342SR.142	19	.232	40	F60341	995S	5100FL
14/4	342SR.144	19	.260	66	-----	-----	5102FL
12/2	342SR.122	19	.266	59	F70095	999S	5000FL

Manufacturer's makes vary. Diameters (Col. B) and weights (Col. C) are conservative and for engineering purposes only. Consult manufacturer's data for individual make & model information.

Column A: Stranded bare copper conductor number of strands.
Column B: Nominal Cable Diameter in inches, for conduit fill calculations.
Column C: Nominal cable weight in pounds per 1000ft.

342SP: 300V Specification 342: Stranded Bare Copper Fire Alarm Cables Shielded
Str BC / Cabled / Fire Alarm / Shielded (S) / Jacket Plenum Rated (P).

Conductor: Stranded Bare Copper. Number of strands per Table 342SP Column A.

Insulation: LS-PVC. Thickness according to UL 1424.

Color Code: Black, Red, White, Green.

Assembly: Twisted together. 2C is 1 pair, 3C is 1 Triad, and 4C is one quad.
Ripcord under the jacket to facilitate removal.

Shielding: Overall shield of aluminum-polyester shielding tape and tinned copper drain wire.

Jacket: LS-PVC. Thickness according to UL 1424.
Color: Red unless specified.

Diameter: Nominal diameter shall not exceed that in table 342SP Column B.

Flame test: Passes NFPA 262 (formerly UL 910).

Voltage: 300V but not printed on the jacket per UL 1424.

Application: Fire Power Limited Circuit Cable per NEC Art. 760 and UL 1424.

Installation: Suitable for use as open cabling in plenum spaces, type FPLP. Also for use in risers and in horizontal runs since rated higher than those requirements.

Temp.: 0C to 60C minimum range, continuous operating ambient temperature.

QA: UL Listed as cable type FPLP. Cable ratings shall be printed on the jacket.
UL file number will be printed on the jacket alongside the gauge size and cable rating.
UL label shall be displayed on every cable package near manufacturer's label.

Packaging: 1000ft or 500ft: spools, reels or boxes.

Appr. Manf: Manufacturer make and model shall be as listed in table 342SP.

Availability: Cable shall be available from local distribution known to project management.

Table 342SP Str BC; Cabled; Fire Alarm; Shielded (S); Plenum (P) Rated.

Const	SDN	A	B	C	TAPPAN	BELDEN-WEST PENN	BELDEN
18/2	342SP.182	7	.162	20	G40178	-----	6300FL
18/4	342SP.184	7	.189	33	G40180	-----	6302FL
16/2	342SP.162	7	.184	28	G50360	60990BS	6200FL
16/4	342SP.164	7	.215	45	-----	-----	6202FL
14/2	342SP.142	19	.220	41	G60337	60992BS	6100FL
12/2	342SP.122	19	.264	61	G70093	60994BS	-----

Manufacturer's makes vary. Diameters (Col. B) and weights (Col. C) are conservative and for engineering purposes only. Consult manufacturer's data for individual make & model information.

Column A: Stranded bare copper conductor number of strands.
Column B: Nominal Cable Diameter in inches, for conduit fill calculations.
Column C: Nominal cable weight in pounds per 1000ft.

343UP: 300V Specification 343: Solid BC NYC LL5 Fire Alarm Cable 150C; Unshielded
Solid Bare Copper / Fire Alarm Cable / NYC LL5 150C / Unshld (U) / Jkt. / Plenum NYC LL5 150C.

Conductor: Solid Bare Copper.

Insulation: Fluoropolymer. Thickness according to UL 1424.

Color Code: Black / Red and White / Green

Assembly: One twisted pair or two twisted pairs. Unshielded.

Shielding: Unshielded.

Jacket: Fluoropolymer. Thickness according to UL 1424. Color: Red, unless specified.

Diameter: Nominal diameter shall not exceed that in table 343UP Column B.

Flame test: Passes NFPA 262 (UL 910 Steiner Tunnel) flame test for use in plenum spaces.

Voltage: 300V but not printed on the jacket per UL 1424.

Application: Fire Power Limited Circuit Cable per NEC Art. 760 and UL 1424.
For use in New York City per Local Law 5.

Installation: Suitable for open cabling use in plenum spaces as FPLP. Also for open cabling use in risers and in horizontal runs since rated higher than those requirements.
For use in New York City per Local Law 5.

Temp.: 0C to 150C.

QA: UL Listed as cable type FPLP. Cable rating(s) shall be printed on the cable or on a tape under the jacket. UL file number will be printed on the jacket or on the tape.
UL label shall be displayed on every cable package near manufacturer's label.

Packaging: 1000ft or 500ft: spools, reels or boxes.

Appr. Manf: Manufacturer make and model shall be as listed in table 343UP.

Availability: Cable shall be available from local distribution known to project management.

Table 343UP Sol BC; Fire Alarm Cable NYC LL5 Compliant 150C; N/S; Plenum Rated.

Const	SDN	CAP	TEMP	B	C	TAPPAN	BELDEN-WEST PENN
18/1pr	343UP.181	2	150C	.191	25	EK4983	LL5182
18/2pr	343UP.182	2	150C	.239	43	EK4984	LL5184
16/1pr	343UP.161	2	150C	.212	32	EK5986	LL5162
16/2pr	343UP.162	2	150C	.262	58	EK5981	LL5164
14/1pr	343UP.141	2	150C	.238	45	EK6981	LL5142
14/2pr	343UP.142	2	150C	.318	78	EK6977	-----
12/1pr	343UP.121	3	150C	.272	62	EK7981	LL5122
12/2pr	343UP.122	2	150C	.438	117	EK7983	-----

Manufacturer's makes vary. Diameters (Col. B) and weights (Col. C) are conservative and for engineering purposes only. Consult manufacturer's data for individual make & model information.

CAP: Capacitance in pF/ft '1' = <16 '2'= 16-24 '3"=25-35 '4'=36-49 '5'= 50 & up
Column B: Cable Diameter in inches.
Column C: Nominal cable weight in pounds per 1000ft

343SP: 300V Specification 343: Solid BC NYC LL5 Fire Alarm Cable 150C; **Shielded**
Solid Bare Copper / Fire Alarm Cable / NYC LL5 150C / Shielded (S) / Jkt. / Plenum. **NYC LL5 150C**

Conductor: Solid Bare Copper.

Insulation: Fluoropolymer. Thickness according to UL 1424.

Color Code: Black / Red and White / Green

Assembly: One twisted pair or two twisted pairs. Shielded overall.

Shielding: Aluminum-polyester tape and tinned copper drain wire, over the cable core.

Jacket: Fluoropolymer. Thickness according to UL 1424. Color: Red, unless specified.

Diameter: Nominal diameter shall not exceed that in table 343SP Column B.

Flame test: Passes NFPA 262 (UL 910 Steiner Tunnel) flame test for use in plenum spaces.

Voltage: 300V but not printed on the jacket per UL 1424.

Application: Fire Power Limited Circuit Cable per NEC Art. 760 and UL 1424.
For use in New York City per Local Law 5..

Installation: Suitable for open cabling use in plenum spaces as FPLP. Also for open cabling use in risers and in horizontal runs since rated higher than those requirements.
For use in New York City per Local Law 5.

Temp.: 0C to 150C.

QA: UL Listed as cable type FPLP. Cable rating(s) shall be printed on the cable or on a tape under the jacket. UL file number will be printed on the jacket or on the tape..
UL label shall be displayed on every cable package near manufacturer's label.

Packaging: 1000ft or 500ft: spools, reels or boxes.

Appr. Manf: Manufacturer make and model shall be as listed in table 343SP.

Availability: Cable shall be available from local distribution known to project management.

Table 343SP Sol BC; Fire Alarm Cable NYC LL5 Compliant 150C; Shld; Plenum Rated.

Const	SDN	CAP	TEMP	B	C	TAPPAN	BELDEN-WEST PENN
18/1pr	343SP.181	3	150C	.194	29	EK4987	LL5182SH
18/2pr	343SP.182	3	150C	.243	45	EK4982	LL5184SH
16/1pr	343SP.161	3	150C	.215	34	EK5983	LL5162SH
16/2pr	343SP.162	3	150C	.267	60	EK5980	LL5164SH
14/1pr	343SP.141	4	150C	.242	47	EK6974	LL5142SH
14/2pr	343SP.142	4	150C	.318	82	EK6975	-----
12/1pr	343SP.121	4	150C	.272	64	EK7980	LL5122SH
12/2pr	343SP.122	4	150C	.438	122	EK7982	-----

Manufacturer's makes vary. Diameters (Col. B) and weights (Col. C) are conservative and for engineering purposes only. Consult manufacturer's data for individual make & model information.

CAP: Capacitance in pF/ft '1' = <16 '2'= 16-24 '3''=25-35 '4'=36-49 '5'= 50 & up
Column B: Cable Diameter in inches.
Column C: Nominal cable weight in pounds per 1000ft

344UP: 300V Specification 344: Solid BC NYC LL5 Fire Alarm Cable 200C; Solid Bare Copper / Fire Alarm Cable / NYC LL5 200C / Unshld (U) / Jkt. / Plenum **Unshielded NYC LL5 200C**

Conductor: Solid Bare Copper.

Insulation: Fluoropolymer. Thickness according to UL 1424.

Color Code: Black / Red and White / Green

Assembly: One twisted pair or two twisted pairs. Unshielded.

Shielding: Unshielded.

Jacket: Fluoropolymer. Thickness according to UL 1424. Color: Red, unless specified.

Diameter: Nominal diameter shall not exceed that in table 344UP Column B.

Flame test: Passes NFPA 262 (UL 910 Steiner Tunnel) flame test for use in plenum spaces.

Voltage: 300V but not printed on the jacket per UL 1424.

Application: Fire Power Limited Circuit Cable per NEC Art. 760 and UL 1424.
For use in New York City per Local Law 5.

Installation: Suitable for open cabling use in plenum spaces as FPLP. Also for open cabling use in risers and in horizontal runs since rated higher than those requirements.
For use in New York City per Local Law 5.

Temp.: 0C to 200C.

QA: UL Listed as cable type FPLP. Cable rating(s) shall be printed on the cable or on a tape under the jacket. UL file number will be printed on the jacket or on the tape..
UL label shall be displayed on every cable package near manufacturer's label.

Packaging: 1000ft or 500ft: spools, reels or boxes.

Appr. Manf: Manufacturer make and model shall be as listed in table 344UP.

Availability: Cable shall be available from local distribution known to project management.

Table 344UP Sol BC; Fire Alarm Cable NYC LL5 Compliant 200C; N/S; Plenum Rated.

Const	SDN	CAP	TEMP	B	C	TAPPAN
18/1pr	344UP.181	2	200C	.191	28	ET4998
18/2pr	344UP.182	2	200C	.230	46	ET4984
16/1pr	344UP.161	2	200C	.212	37	ET5979
16/2pr	344UP.162	2	200C	.260	62	ET5977
14/1pr	344UP.141	2	200C	.238	49	ET6986
14/2pr	344UP.142	2	200C	.318	88	ET6985
12/1pr	344UP.121	2	200C	.272	69	ET7982

CAP: Capacitance in pF/ft '1' = <16 '2'= 16-24 '3"=25-35 '4'=36-49 '5'= 50 & up
Column B: Cable Diameter in inches.
Column C: Nominal cable weight in pounds per 1000ft

344SP: 300V Specification 344: Solid BC NYC LL5 Fire Alarm Cable 200C; Shielded
Solid Bare Copper / Fire Alarm Cable / NYC LL5 200C / Shielded (S) / Jkt. / Plenum. NYC LL5 200C

Conductor: Solid Bare Copper.

Insulation: Fluoropolymer. Thickness according to UL 1424.

Color Code: Black / Red and White / Green

Assembly: One twisted pair or two twisted pairs. Shielded overall.

Shielding: Aluminum-polyester tape and tinned copper drain wire, over the cable core.

Jacket: Fluoropolymer. Thickness according to UL 1424. Color: Red, unless specified.

Diameter: Nominal diameter shall not exceed that in table 344SP Column B.

Flame test: Passes NFPA 262 (UL 910 Steiner Tunnel) flame test for use in plenum spaces.

Voltage: 300V but not printed on the jacket per UL 1424.

Application: Fire Power Limited Circuit Cable per NEC Art. 760 and UL 1424.
For use in New York City per Local Law 5..

Installation: Suitable for open cabling use in plenum spaces as FPLP. Also for open cabling use in risers and in horizontal runs since rated higher than those requirements.
For use in New York City per Local Law 5.

Temp.: 0C to 200C.

QA: UL Listed as cable type FPLP. Cable rating(s) shall be printed on the cable or on a tape under the jacket. UL file number will be printed on the jacket or on the tape..
UL label shall be displayed on every cable package near manufacturer's label.

Packaging: 1000ft or 500ft: spools, reels or boxes.

Appr. Manf: Manufacturer make and model shall be as listed in table 344SP.

Availability: Cable shall be available from local distribution known to project management.

Table 344SP Sol BC; Fire Alarm Cable NYC LL5 Compliant 200C; Shld; Plenum Rated.

Const	SDN	CAP	TEMP	B	C	TAPPAN
18/1pr	344SP.181	2	200C	.191	30	ET4986
18/2pr	344SP.182	2	200C	.230	49	ET4985
16/1pr	344SP.161	3	200C	.212	39	ET5978
16/2pr	344SP.162	3	200C	.260	65	ET5976
14/1pr	344SP.141	3	200C	.238	53	ET6983
14/2pr	344SP.142	3	200C	.318	92	ET6984
12/1pr	344SP.121	4	200C	.272	72	ET7985

CAP: Capacitance in pF/ft '1' = <16 '2'= 16-24 '3''=25-35 '4'=36-49 '5'= 50 & up
Column B: Cable Diameter in inches.
Column C: Nominal cable weight in pounds per 1000ft

345UCI: 300V Specification 345: Solid BC Fire Alarm Circuit Integrity; Unshielded
Solid Bare Copper / Fire Alarm Cable / FPLR-CI / Unshielded (U) / Jacket UL 2196 Open Cabling

Conductor: Solid or Stranded Bare Copper per Table 345UCI Column A.

Insulation: Ceramifiable Silicone Rubber, or composite of mica tape and ceramifiable silicone rubber.
Color Code: 2C cable is Black, Red. 3 or more conductor cable is black with printed numbers.

Assembly: Twisted together so that a 2 conductor cable is also considered 1 pair. 4 conductor cable is twisted but may be called 2 pair even when not assembled as 2 pair cable.

Shielding: Unshielded.
Barrier Tape: An overall flame barrier tape under the overall jacket.

Jacket: Flame retardant polyethylene.
Diameter: Nominal diameter shall not exceed that in table 345UCI.
Flame test: Passes UL 1666 for use as open cabling in risers. Passes UL 2196 as open cabling and listed in the UL Fire Resistance Directory as Fire Resistive Cable (FHJR.)
Voltage: For low voltage systems. As prescribed by the cable manufacturer's FHJR listing..
Application: Fire Power Limited Circuit Cable per NEC Art. 760 and UL 1424. For use where survivability from attack by fire, 2 Hour Circuit Integrity through one or more fire zones is specified.

Installation: Suitable as open cabling installed as Fire Resistive Cable (FHJR.)
Temp.: 0C to 60C continuous (UL 1424); In a fire scenario to 1000F in 5 minutes and to 1850F after 15 minutes and up to for 2 hours (UL 2196.) After which the cable should be replaced.
QA: UL Listed as cable type FPLR-CI. Cable rating(s) shall be printed on the jacket.
UL file number will be printed on the jacket alongside AWG size and flame rating.
UL label shall be displayed on every cable package near manufacturer's label.
Packaging: 1000ft reels.
Appr. Manf: Manufacturer make and model shall be as listed in table 345UCI.
Availability: Cable shall be available from local distribution known to project management.

Table 345UCI BC; Fire Alarm Circuit Integrity Cable; N/S; UL 2196 Open Cabling Riser Rated.

Const	SDN	A	CAP	B	C	TAPPAN	PRYSMIAN DRAKA	BELDEN or WEST PENN
18/2	345UCI.182	1	2	.32	50	SC4001	CIFPLR-2C/18NS	5320UM
18/4	345UCI.184	1	2	.37	77	SC4002	CIFPLR-4C/18NS	5322UM
16/2	345UCI.162	1	2	.33	60	SC5001	CIFPLR-2C/16NS	5220UM
16/4	345UCI.164	1	2	.39	96	SC5002	CIFPLR-4C/16NS	5222UM
14/2	345UCI.142A	1	2	.37	76	SC6006	------	5120UM
	345UCI.142B	Str	2	.38	76	SC6001	CIFPLR-2C/14NS	------
14/4	345UCI.144A	1	2	.42	115	-----	------	5122UM
	345UCI.144B	Str	2	.44	126	SC6002	CIFPLR-4C/14NS	------
12/2	345UCI.122A	1	3	.40	98	-----	------	5020UM
	345UCI.122B	Str	3	.41	99	SC7001	CIFPLR-2C/12NS	-----
12/4	345UCI.124A	1	2	.46	157	-----	------	5022UM
	345UCI.124B	Str	2	.49	170	SC7002	CIFPLR-4C/12NS	-----

Manufacturer's makes vary. Diameters (Col. B) and weights (Col. C) are conservative and for engineering purposes only. Consult manufacturer's data for individual make & model information.

CAP: Capacitance in pF/ft '1' = <16 '2'= 16-24 '3''=25-35 '4'=36-49 '5'= 50 & up
Column B: Cable Diameter in inches.
Column C: Nominal cable weight in pounds per 1000ft

345SCI: 300V Specification 345: Solid BC Fire Alarm Circuit Integrity; Shielded
Solid Bare Copper / Fire Alarm Cable / FPLR-CI / Shielded (S) / Jacket UL 2196 Open cabling.

Conductor: Solid or Stranded Bare Copper per Table 345SCI Column A.

Insulation: Ceramifiable Silicone Rubber, or composite of mica tape and ceramifiable silicone rubber.

Color Code: 2C cable is Black, Red. 3 or more conductor cable is black with printed numbers.

Assembly: Twisted together so that a 2 conductor cable is also considered 1 pair. 4 conductor cable is twisted but may be called 2 pair even when not assembled as 2 pair cable.

Shielding: Overall shield of metallic foil plus a drain wire or .005" metallic tape; or per Mfg. UL listing.

Barrier Tape: An overall flame barrier tape under the overall jacket.

Jacket: Flame retardant polyethylene.

Diameter: Nominal diameter shall not exceed that in table 345SCI.

Flame test: Passes UL 1666 for use as open cabling in risers. Passes UL 2196 as open cabling and listed in the UL Fire Resistance Directory as Fire Resistive Cable (FHJR.)

Voltage: For low voltage systems. As prescribed by the cable manufacturer's FHJR listing..

Application: Fire Power Limited Circuit Cable per NEC Art. 760 and UL 1424. For use where survivability from attack by fire, 2 Hour Circuit Integrity through one or more fire zones is specified.

Installation: Suitable as open cabling installed as Fire Resistive Cable (FHJR.)

Temp.: 0C to 60C continuous (UL 1424); In a fire scenario to 1000F in 5 minutes and to 1850F after 15 minutes and up to for 2 hours (UL 2196.) After which the cable should be replaced.

QA: UL Listed as cable type FPLR-CI. Cable rating(s) shall be printed on the jacket.
UL file number will be printed on the jacket alongside AWG size and flame rating.
UL label shall be displayed on every cable package near manufacturer's label.

Packaging: 1000ft reels.

Appr. Manf: Manufacturer make and model shall be as listed in table 345SCI.

Availability: Cable shall be available from local distribution known to project management.

Table 345SCI BC; Fire Alarm Circuit Integrity Cable; Shld; UL 2196 Open cabling Rated.

Const	SDN	A	CAP	B	C	TAPPAN	PRYSMIAN DRAKA	BELDEN or WEST PENN
18/2	345SCI.182	1	3	.33	57	SC4009	CIFPLR-2C/18OS	5320FM
18/4	345SCI.184	1	3	.38	84	SC4010	CIFPLR-4C/18OS	5322FM
16/2	345SCI.162	1	3	.34	67	SC5007	CIFPLR-2C/16OS	5220FM
16/4	345SCI.164	1	3	.58	120	SC5008	CIFPLR-4C/16OS	5222FM
14/2	345SCI.142A	1	4	.39	84	SC6011	-----	5120FM
	345SCI.142B	Str	4	.39	84	SC6007	CIFPLR-2C/14OS	5100FM
14/4	345SCI.144A	1	3	.42	125	-----	-----	5122FM
	345SCI.144B	Str	4	.67	153	SC6008	CIFPLR-4C/14OS	5102FM
12/2	345SCI.122	Str	4	.43	109	SC7007	CIFPLR-2C/12OS	5000FM
12/4	345SCI.124	Str	4	.49	182	SC7008	CIFPLR-4C/12OS	5002FM

Manufacturer's makes vary. Diameters (Col. B) and weights (Col. C) are conservative and for engineering purposes only. Consult manufacturer's data for individual make & model information.

CAP: Capacitance in pF/ft '1' = <16 '2'= 16-24 '3''=25-35 '4'=36-49 '5'= 50 & up
Column B: Cable Diameter in inches.
Column C: Nominal cable weight in pounds per 1000ft

345UCIC: 300V Specification 345: Solid BC Fire Alarm Circuit Integrity; Unshielded
Solid Bare Copper / Fire Alarm Cable / FPLR "CIC" / Unshielded (U) / Jacket UL 2196 In-Conduit.

Conductor: Solid or Stranded Bare Copper per Table 345UCIC Column A.

Insulation: Ceramifiable Silicone Rubber, or composite of mica tape and ceramifiable silicone rubber.

Color Code: 2C cable is Black, Red. 3 or more conductor cable is black with printed numbers.

Assembly: Twisted together so that a 2 conductor cable is also considered 1 pair. 4 conductor cable is twisted but may be called 2 pair even when not assembled as 2 pair cable.

Shielding: Unshielded.

Barrier Tape: An overall flame barrier tape under the overall jacket.

Jacket: Flame retardant polyethylene.

Diameter: Nominal diameter shall not exceed that in table 345UCIC.

Flame test: UL 1581 (FPL) or UL 1666 (FPLR.) Listed in an FHIT Fire Resistive System of cable and conduit having passed UL 2196 for conduit installations.

Voltage: For low voltage systems. As prescribed by the cable manufacturer's FHJR listing..

Application: Fire Power Limited Circuit Cable per NEC Art. 760 and UL 1424. For use where survivability from attack by fire, 2 Hour Circuit Integrity through one or more fire zones is specified.

Installation: Suitable for use in UL listed cable in conduit Fire Resistive System (FHIT) according to Mfg.

Temp.: 0C to 60C continuous (UL 1424); In a fire scenario: to 1000F in 5 minutes and to 1850F after 15 minutes and up to for 2 hours (UL 2196.) After which the cable should be replaced.

QA: UL Listed as cable type FPL or FPLR. Cable rating(s) shall be printed on the jacket.
UL file number will be printed on the jacket alongside AWG size and flame rating.
UL label shall be displayed on every cable package near manufacturer's label.
Included in an FHIT UL Fire Resistive system of cable & conduit.

Packaging: 1000ft reels.

Appr. Manf: Manufacturer make and model shall be as listed in table 345UCI.

Availability: Cable shall be available from local distribution known to project management.

Table 345UCIC BC; Fire Alarm Circuit Integrity Cable; N/S; UL 2196 In-Conduit Rated (FHIT)

Const	SDN	A	CAP	B	C	TAPPAN	PRYSMIAN DRAKA	BELDEN or WEST PENN
18/2	345UCIC.182	1	2	.36	50	SC4001	CICFPL-2C/18NS	5320UZ
18/4	345UCIC.184	1	2	.38	77	SC4002	CICFPL-4C/18NS	-----
16/2	345UCIC.162	1	2	.35	60	SC5001	CICFPL-2C/16NS	5220UZ
16/4	345UCIC.164	1	2	.39	96	SC5002	CICFPL-4C/16NS	-----
14/2	345UCIC.142A	1	2	.38	76	SC6006		5120UZ
	345UCIC.142B	Str	2	.40	76	SC6001	CICFPL-2C/14NS	5100UZ
14/4	345UCIC.144B	Str	2	.44	126	SC6002	CICFPL-4C/14NS	-----
12/2	345UCIC.122	Str	2	.44	99	SC7001	CICFPL-2C/12NS	5000UZ
12/4	345UCIC.124	Str	2	.45	145	-----	CICFPL-4C/12NS	-----

Manufacturer's makes vary. Diameters (Col. B) and weights (Col. C) are conservative and for engineering purposes only. Consult manufacturer's data for individual make & model information.

CAP: Capacitance in pF/ft '1' = <16 '2'= 16-24 '3"=25-35 '4'=36-49 '5'= 50 & up
Column B: Cable Diameter in inches.
Column C: Nominal cable weight in pounds per 1000ft

345SCIC: 300V Specification 345: Solid BC Fire Alarm Circuit Integrity; Shielded
Solid Bare Copper / Fire Alarm Cable / FPLR "CIC" / Shielded (S) / Jacket UL 2196 In-Conduit.

Conductor: Solid or Stranded Bare Copper per Table 345SCIC Column A.

Insulation: Ceramifiable Silicone Rubber, or composite of mica tape and ceramifiable silicone rubber.

Color Code: 2C cable is Black, Red. 3 or more conductor cable is black with printed numbers.

Assembly: Twisted together so that a 2 conductor cable is also considered 1 pair. 4 conductor cable is twisted but may be called 2 pair even when not assembled as 2 pair cable.

Shielding: Overall shield of metallic foil plus a drain wire or .005" metallic tape; or per Mfg. UL listing.

Barrier Tape: An overall flame barrier tape under the overall jacket.

Jacket: Flame retardant polyethylene.

Diameter: Nominal diameter shall not exceed that in table 345UCIC.

Flame test: UL 1581 (FPL) or UL 1666 (FPLR.) Listed in an FHIT Fire Resistive System of cable and conduit having passed UL 2196 for conduit installations.

Voltage: For low voltage systems. As prescribed by the cable manufacturer's FHJR listing..

Application: Fire Power Limited Circuit Cable per NEC Art. 760 and UL 1424. For use where survivability from attack by fire, 2 Hour Circuit Integrity through one or more fire zones is specified.

Installation: Suitable for use in UL listed cable in conduit Fire Resistive System (FHIT) according to Mfg.

Temp.: 0C to 60C continuous (UL 1424); In a fire scenario: to 1000F in 5 minutes and to 1850F after 15 minutes and up to for 2 hours (UL 2196.) After which the cable should be replaced.

QA: UL Listed as cable type FPL or FPLR. Cable rating(s) shall be printed on the jacket.
UL file number will be printed on the jacket alongside AWG size and flame rating.
UL label shall be displayed on every cable package near manufacturer's label.
Included in an FHIT UL Fire Resistive system of cable & conduit.

Packaging: 1000ft reels.

Appr. Manf: Manufacturer make and model shall be as listed in table 345UCI.

Availability: Cable shall be available from local distribution known to project management

Table 345SCIC BC; Fire Alarm Circuit Integrity Cable; Shld; UL 2196 In-Conduit Rated (FHIT.)

Const	SDN	A	CAP	B	C	TAPPAN	PRYSMIAN DRAKA	BELDEN or WEST PENN
18/2	345SCIC.182	1	3	.37	57	SC4013	CICFPL-2C/18OS	5320FZ
18/4	345SCIC.184	1	3	.38	84	SC4014	CICFPL-4C/18OS	-----
16/2	345SCIC.162	1	3	.36	67	SC5007	CICFPL-2C/16OS	5220FZ
16/4	345SCIC.164	1	3	.58	150	SC5008	CICFPL-4C/16OS	5222FZ
14/2	345SCIC.142A	1	2	.39	84	SC6011	-----	5120FZ
	345SCIC.142B	Str	4	.41	88	SC6007	CICFPL-2C/14OS	5100FZ
14/4	345SCIC.144A	1	2	.44	133	-----	-----	5122FZ
	345SCIC.144B	Str	4	.67	153	SC6008	CICFPL-4C/14OS	-----
12/2	345SCIC.122	Str	4	.44	114	SC7007	CICFPL-2C/12OS	5000FZ
12/4	345SCIC.124	Str	4	.49	182	SC7008	CICFPL-4C/12OS	-----

Manufacturer's makes vary. Diameters (Col. B) and weights (Col. C) are conservative and for engineering purposes only. Consult manufacturer's data for individual make & model information.

CAP: Capacitance in pF/ft '1' = <16 '2'= 16-24 '3"=25-35 '4'=36-49 '5'= 50 & up
Column B: Cable Diameter in inches.
Column C: Nominal cable weight in pounds per 1000ft

MASTER TABLE OF PUBLIC CABLE 1.0 PRODUCTS

PERIODIC TABLE OF PARTIAL CONTENTS – Detail for Coaxial Cable Types

Comminity Antenna Television Coaxial Cable

Specification 351CA

Cable Type	Installation	Spec / Table	Spec. Page	Table Page
Community Antenna CATV	Horizontal	351CAH	150	150
	Plenum	351CAP	151	151

Closed Circuit Television Coaxial Cable / Surveillance Camera

Specification 352CC

Cable Type	Installation	Spec / Table	Spec. Page	Table Page
Closed Circuit TV CCTV	Horizontal	352CCH	152	152
	Plenum	352CCP	153	153

CCTV Coaxial Cable + Power Pair

Specification 353CC

Cable Type	Installation	Spec / Table	Spec. Page	Table Page
CCTV + Power Pair	Horizontal	353CCH	154	154
	Plenum	353CCP	155	155

HD-SDI Coaxial Cable

Specification 354SD

Cable Type	Installation	Spec / Table	Spec. Page	Table Page
HD-SDI	Riser	354SDR	156	156
	Plenum	354SDP	157	157

RGBHV Component Video Mini Hi-Resolution Coaxial Cable

Specification 355RGB

Cable Type	Installation	Spec / Table	Spec. Page	Table Page
RGBHV	Horiz. / Riser	355RGBH	158	158
	Plenum	355RGBP	159	159

50Ω RF Transmission Coaxial Cable

Specification 356RFX

Cable Type	Installation	Spec / Table	Spec. Page	Table Page
50Ω RF Transmission	Limited / Horiz	356RFX	160	160

351CAH: 300V Specification 351: CATV Coax Cable **CATV - H**
Solid Conductor / Foam PE Dielectric / Dual or Quad CATV Shield / PVC Jkt **For Horizontal Use.**

Conductor: Solid center conductor, copper-clad-steel (CCS) or Solid Bare Copper (BC)

Insulation: Foam Polyethylene per UL 1655 and/or UL 13 or UL 444; and as required for 75 Ohm impedance; and as required for standardized connectors to fit.

Shielding: CATV shielding: either: (a) Dual Shield: Alum.-Poly-Alum Tape + 60% Aluminum Braid or (b) Quad Shield Alum.-Poly-Alum Tape+60% Aluminum Braid+ Alum.-Poly-Alum Tape+40% Aluminum Braid or (c) Alum.-Poly-Alum Tape + 90% Aluminum Braid; as indicated in table.

Jacket: Black PVC Jacket. (Other colors as may be specified on drawings.)
Diameter: Per Table 351CAH - Column B.
Flame test: Passes UL 1685 Vertical Tray / CSA FT1.
Voltage: No voltage rating (CATV only.) Or 150V or 300V rated. Cables listed as CL2 (only) are rated 150V. Others are rated 300V. Voltage rating is not printed on the jacket UL 13 and UL 444. Cables rated CATV (only) do not have a (UL) voltage rating; per UL 1655.
Application: Broadband video, CATV, internet, and telephone signal where an F connector is provided. Cables with a solid BC center conductor are suitable for DBS satellite systems.
Installation: Suitable open wiring / open cabling use in concealed horizontal runs as Type CATV; NEC 820.
Temp.: 0C to 60C minimum range; continuous operating ambient temperature.
QA: UL Listed as cable type CATV and/or CL2, CL3, or CM. Cable rating(s) printed on the jacket. UL file number will be printed on the jacket alongside AWG and cable type. UL label shall be displayed on every cable package near manufacturer's label.
Packaging: 1000ft or 500ft: spools or reels.
Appr. Manf: Manufacturer make and model shall be as selected from table 351CAH.
Availability: Cable shall be available from local distribution known to project management.

Table 351CAH: CATV Coax; Solid conductor / Foam PE / Shielding / PVC Jacket. Horizontal Use

Const	SDN	A	Shield	B	C	TAPPAN	General Cable	Belden: WEST PENN	BELDEN
RG 6	351CAH.181DA	CCS	Dual	.270	35	C40055	C5775	6100	9116
	351CAH.1819A	CCS	Foil+90%Al	.275	36	-----	C5776	-----	-----
2 x RG 6	351CAH.182DA	CCS	Dual	.275x.595	70	C40076	C5822	6150	9077
RG 6	351CAH.181QA	CCS	Quad	.298	42	C40065	C5785	6300	1189A
2 x RG 6	351CAH.182QA	CCS	Quad	.305x.620	85	C40168	-----	6500	----
RG 11*	351CAH.141DA	CCS	Dual	.405	71	C60008	C5039	1100	1523A
	351CAH.141QA	CCS	Quad	.407	72	C60006	C5044	----	1617A
RG 6	351CAH.181DB	BC	Dual	.270	36	C40008	-----	841	5339B5
	351CAH.1819B	BC	Foil+90%Al	.270	36	C40166	-----	-----	-----
	351CAH.181QB	BC	Quad	.298	36	C40007	-----	Q841	5339Q5
2 x RG 6	351CAH.182QB	BC	Quad	.305x.620	85	C40126	-----	----	----
RG 11*	351CAH.141DB	BC	Dual	.405	74	C60009	----	821	----
	351CAH.141QB	BC	Quad	.405	68	C60007	-----	Q821	----

Manufacturer's makes vary. Diameters (Col. B) and weights (Col. C) are conservative and for engineering purposes only. Consult manufacturer's data for individual make & model information.

Column A: Conductor Material
Column B: Approximate Cable Diameter in inches. (Cable sized to fit the connectors selected.)
Column C: Approximate - nominal cable weight in pounds per 1000ft

* See note at the bottom of table 351CAP (next page.)

351CAP: 300V Specification 351: CATV Coax Cable **CATV - P**
Solid Conductor / Foam FEP Dielectric / CATV Shielding / Suitable Jacket **For Plenum Use.**

Conductor: Solid center conductor, copper-clad-steel (CCS) or Solid Bare Copper (BC)

Insulation: Foam FEP per UL 1655 and/or UL 13 or UL 444; and as required for 75 Ohm impedance; and as required for standardized connectors to fit.

Shielding: Either: (a) Dual Shield: Alum.-Poly-Alum Tape + 60% Aluminum Braid or (b) Quad Shield: Alum.-Poly-Alum Tape+60% Aluminum Braid+ Alum.-Poly-Alum Tape+40% Aluminum Braid or (c) Alum.-Poly-Alum Tape + 90% Aluminum Braid; as indicated in table 351CAP.

Jacket: White or Natural LS-PVC or fluoropolymer jacket. (Other colors as may be specified.)

Diameter: Per Table 351CAP - Column B.

Flame test: Passes NFPA 262 (UL 910 Steiner Tunnel) flame test for use in plenum spaces.

Voltage: No voltage rating (CATVP only.) Or 150V or 300V rated. Cables listed as CL2P (only) are rated 150V. Others are rated 300V. Voltage rating is not printed on the jacket per UL 13 and UL 444. Cables rated CATVP (only) do not have a (UL) voltage rating; per UL 1655.

Application: Broadband video, CATV, internet, and telephone signal where an F connector is provided. Cables with a solid BC center conductor are suitable for DBS satellite systems.

Installation: Suitable for open cabling use in all runs including plenum spaces as Type CATV-P. NEC 820.

Temp.: 0C to 60C minimum range; continuous operating ambient temperature.

QA: UL Listed as cable type CATV-P and/or CL2P, CL3P, or CMP; rating(s) printed on the jacket. UL file number will be printed on the jacket alongside AWG and cable type. UL label shall be displayed on every cable package near manufacturer's label.

Packaging: 1000ft or 500ft: spools or reels.

Appr. Manf: Manufacturer make and model shall be as selected from table 351CAP.

Availability: Cable shall be available from local distribution known to project management.

Table 351CAP: CATV Coax; Sol. Cond / Foam FEP / Shielding / Appropriate Jacket. Plenum Use

Const	SDN	A	Shield	B	C	TAPPAN	General Cable	Belden: WEST PENN	BELDEN
RG 6	351CAP.181DA	CCS	Dual	.235	25	D40018	C3524	256100	9116P
	351CAP.181QA	CCS	Quad	.306	39	D40019	C3525	256300	1189AP
RG 11*	351CAP.141DA	CCS	Dual	.348	57	D90011	C3528	-----	1523AP
	351CAP.141QA	CCS	Quad	.387	106	D60004	C3529	-----	1153A**
RG 6	351CAP.181DB	BC	Dual	.230	26	D40016	-----	-----	-----
	351CAP.1819B	BC	Foil+90%Al	.244	28	D40046	-----	25841	633938
	351CAP.181QB	BC	Quad	.273	30	D40017	-----	25Q841	-----
RG 11*	351CAP.141DB	BC	Dual	.352	59	D90011	-----	25821	6139B8
	351CAP.141QB	BC	Quad	.387	62	D90010	-----	25Q821	-----

Manufacturer's makes vary. Diameters (Col. B) and weights (Col. C) are conservative and for engineering purposes only. Consult manufacturer's data for individual make & model information.

Column A: Conductor Material
Column B: Approximate Cable Diameter in inches. (Cable sized to fit the connectors selected.)
Column C: Approximate - nominal cable weight in pounds per 1000ft

* Manufacturers listing as cable CL2P (only) – the voltage rating is 150V. Manufacturers listing as CATV-P (only) – the cable does not have a (UL) voltage rating. Cables rated CL3P or CMP are rated 300V (UL).

** 1153A is made with two tinned copper braids rather than aluminum braid wire.

352CCH: 300V Specification 352: Solid BC CCTV Coax Cable CCTV - H
Solid Bare Copper / Foam PE Dielectric / 75Ω / 95% Bare Copper Braid / PVC Jkt For Horizontal Use.

Conductor: Solid Bare Copper

Insulation: Foam Polyethylene per UL 13 or UL 444; and as required for 75 Ohm impedance; and as required for standardized connectors to fit.

Shielding: 95% Bare Copper Braid. Cable with copper-clad-aluminum braid or alloy braid is not permitted.

Jacket: Black PVC Jacket. (Other colors as may be specified on drawings.)

Diameter: Per Table 352CCH - Column B.

Flame test: Passes UL 1685 Vertical Tray / CSA FT1.

Voltage: 150V or 300V rated. Cables listed as CL2 (only) are rated 150V. Others are rated 300V. Voltage rating is not printed on the jacket per UL 13 and UL 444

Application: Baseband video; CCTV, HD-CCTV. All closed circuit surveillance video systems where a BNC connection it provided.

Installation: Suitable for open cabling use in concealed horizontal runs as Type CL2, CL3, or CM.

Temp.: 0C to 60C; minimum range, continuous operating ambient temperature.

QA: UL Listed as cable type CL2, CL3, or CM. Cable rating(s) shall be printed on the jacket. UL file number will be printed on the jacket alongside AWG and cable type information. UL label shall be displayed on every cable package near manufacturer's label.

Packaging: 1000ft or 500ft: spools, reels or boxes.

Appr. Manf: Manufacturer make and model shall be as selected from table 352CCH.

Availability: Cable shall be available from local distribution known to project management.

Table 352CCH: CCTV Coax; BC / Foam PE / 75Ω / BC Braid / PVC Jacket. Horizontal Use

Const	SDN	A	B	C	TAPPAN	General Cable	Belden: WEST PENN	BELDEN
Mini 25	352CCH.251	1	.146	16	C90005	-----	825	573945
RG 59	352CCH.201	1	.240	36	C30022	C1142	815	543945
RG 6	352CCH.181	1	.270	44	C40009	-----	806	533945
RG 11*	352CCH.141*	1	.405	88	C60003	395058	811	513945

Manufacturer's makes vary. Diameters (Col. B) and weights (Col. C) are conservative and for engineering purposes only. Consult manufacturer's data for individual make & model information.

Column A: Stranding (Number of Strands)
Column B: Approximate Cable Diameter in inches. (Cable sized to fit the connectors selected.)
Column C: Nominal cable weight in pounds per 1000ft. (Approximate.)

* Manufacturers listing as cable CL2P (only) – the voltage rating is 150V

352CCP: 300V Specification 352: Solid BC CCTV Coax Cable **CCTV - P**
Solid Bare Copper / Foam FEP / 75Ω / 95% Bare Copper Braid / Suitable Jacket **For Plenum Use.**

Conductor: Solid Bare Copper

Insulation: Foam fluoropolymer per UL 13 or UL 444 and as required for 75 Ohm Impedance; and as required for standardized connectors to fit.

Shielding: 95% Bare Copper Braid. Cable with copper-clad-aluminum braid or alloy braid is not permitted.

Jacket: White or Natural LS-PVC or fluoropolymer jacket.
(Other colors as may be specified on drawings.)

Diameter: Per Table 352CCP - Column B.

Flame test: Passes NFPA 262 (UL 910 Steiner Tunnel) flame test for use in plenum spaces.

Voltage: 150V or 300V rated. Cables listed as CL2P (only) are rated 150V. Others are rated 300V.
Voltage rating is not printed on the jacket per UL 13 and UL 444

Application: Baseband video; CCTV, HD-CCTV. All closed circuit surveillance video systems where a BNC connection it provided.

Installation: Suitable for open cabling use in plenum runs as Type CL2P, CL3P, or CMP. Also suitable for use in concealed horizontal runs and in risers, since rated higher than those requirements.

Temp.: 0C to 60C; minimum range, continuous operating ambient temperature.

QA: UL Listed as cable type CL2P, CL3P, or CMP. Cable rating(s) shall be printed on the jacket.
UL file number will be printed on the jacket alongside AWG and cable type information.
UL label shall be displayed on every cable package near manufacturer's label.

Packaging: 1000ft or 500ft: spools, reels or boxes.

Appr. Manf: Manufacturer make and model shall be as selected from table 352CCP.

Availability: Cable shall be available from local distribution known to project management.

Table 352CCP: CCTV Coax; BC / Foam FEP / 75Ω / BC Braid / Suitable Jacket. Plenum Use

Const	SDN	A	B	C	TAPPAN	General Cable	Belden: WEST PENN	BELDEN
Mini 25	352CCP.251	1	.146	16	D90002	-----	25825	673948
RG 59	352CCP.201	1	.207	28	D30009	495028	25815	643948
RG 6	352CCP.181	1	.236	36	D40012	495035	25806	633948
RG 11*	352CCP.141*	1	.352	75	D60001	495015	25811	613948

Manufacturer's makes vary. Diameters (Col. B) and weights (Col. C) are conservative and for engineering purposes only. Consult manufacturer's data for individual make & model information.

Column A: Stranding (Number of Strands)
Column B: Approximate Cable Diameter in inches. (Cable sized to fit the connectors selected.)
Column C: Nominal cable weight in pounds per 1000ft. (Approximate.)

* Manufacturers listing as cable CL2P (only) – the voltage rating is 150V

353CCH: 300V Specification 353: Solid BC CCTV Coax + Power CCTV+Pr - H
Solid Bare Copper / Foam PE / 75Ω / 95% BC Braid / PVC / w-Siamese Pair / PVC For Horizontal Use.

Conductor: Solid Bare Copper (Coax) & Stranded Bare Copper (Low Voltage Power Pair)

Insulation: Coax: Foam Polyethylene per UL 13 or UL 444 and as required for 75 Ohm Impedance, and so as to fit with standard connectors.
Low Voltage Power Pair: PVC per UL 13 or UL 444.

Shielding: On coax only: 95% Bare Copper Braid. / Copper-clad-AL braid or alloy braid are not permitted.

Jacket: Black PVC Jacket – covering the Coax and the low-voltage power pairs separately but contiguously in Siamese configuration. (Other colors as may be specified on drawings.)

Dimensions: Per Table 353CCH - Column B.

Flame test: Passes UL 1685 Vertical Tray / CSA FT1.

Voltage: 150V or 300V rated. Cables listed as CL2 (only) are rated 150V. Others are rated 300V.
Voltage rating is not printed on the jacket per UL 13 and UL 444

Application: Baseband video; CCTV, HD-CCTV. All closed circuit surveillance video systems where a BNC connection it provided; where additional low voltage power is required.

Installation: Suitable for open cabling use in concealed horizontal runs as Type CL2, CL3, or CM.

Temp.: 0C to 60C; minimum range, continuous operating ambient temperature.

QA: UL Listed as cable type CL2, CL3, or CM. Cable rating(s) shall be printed on the jacket.
UL file number will be printed on the jacket alongside AWG and cable type information.
UL label shall be displayed on every cable package near manufacturer's label.

Packaging: 1000ft or 500ft: spools, reels or boxes.

Appr. Manf: Manufacturer make and model shall be as selected from table 353CCH.

Availability: Cable shall be available from local distribution known to project management.

Table 353CCH: 75Ω CCTV Coax + Low Voltage Power Pair. Horizontal Use

Const	Power Pair	SDN	B	C	TAPPAN	General Cable	Belden: WEST PENN	BELDEN
Mini 25	1 pr 20	353CCH.2520	.151 x .332	29	-----	-----	-----	579945
RG 59	1 pr 18	353CCH.2018	.240 x .465	60	H90369	C8028	2815B	549945
RG 6	1 pr 18	353CCH.1818	.270 x .520	70	H90767	C8029	-----	539945
RG 6	1 pr 16	353CCH.1816	.280 x .580	74	H90454	-----	2806B	-----

Manufacturer's makes vary. Diameters (Col. B) and weights (Col. C) are conservative and for engineering purposes only. Consult manufacturer's data for individual make & model information.

Column A: Stranding (Number of Strands)
Column B: Approximate Cable Diameter in inches. (Cable sized to fit the connectors selected.)
Column C: Approximate - nominal cable weight in pounds per 1000ft

353CCP: 300V Specification 353: Solid BC CCTV Coax + Power **CCTV+Pr - P**
Solid Bare Copper / Foam FEP / 75Ω / 95% BC Braid / LS-PVC / w-Siamese Pair **For Plenum Use.**

<u>Conductor</u>: Solid Bare Copper (Coax) & Stranded Bare Copper (Low Voltage Power Pair)

<u>Insulation</u>: Coax: Foam Fluoropolymer per UL 13 or UL 444 and as required for 75 Ohm Impedance; and as required to fit with standard connectors.
Low Voltage Power Pair: LS-PVC per UL 13 or UL 444.

<u>Shielding</u>: On coax only: 95% Bare Copper Braid. / Copper-clad-AL braid or alloy braid are not permitted.

<u>Jacket</u>: White or natural LS- PVC Jacket – covering the Coax and the low-voltage power pairs separately but contiguously in Siamese configuration. (Other colors as specified on drawings.)

<u>Dimensions:</u> Per Table 353CCP - Column B.

<u>Flame test</u>: Passes NFPA 262 (previously UL 910) / CSA FT 6 for use in plenum spaces.

<u>Voltage</u>: 150V or 300V rated. Cables listed as CL2P (only) are rated 150V. Others are rated 300V. Voltage rating is not printed on the jacket per UL 13 and UL 444

<u>Application</u>: Baseband video; CCTV, HD-CCTV. All closed circuit surveillance video systems where a BNC connection it provided; where additional low voltage power is required.

<u>Installation</u>: Suitable for open cabling use in plenum runs as Type CL2P, CL3P, or CMP. Also suitable for use in concealed horizontal runs and in risers, since rated higher.

<u>Temp.</u>: 0C to 60C; minimum range, continuous operating ambient temperature.

<u>QA</u>: UL Listed as cable type CL2P, CL3P, or CMP. Cable rating(s) shall be printed on the jacket. UL file number will be printed on the jacket alongside AWG and cable type information. UL label shall be displayed on every cable package near manufacturer's label.

<u>Packaging</u>: 1000ft or 500ft: spools, reels or boxes.

<u>Appr. Manf</u>: Manufacturer make and model shall be as selected from table 353CCP.

<u>Availability</u>: Cable shall be available from local distribution known to project management.

Table 353CCP: 75Ω CCTV Coax + Low Voltage Power Pair. Plenum Use

Const	Power Pair	SDN	B	C	TAPPAN	Belden: WEST PENN	BELDEN
Mini 25	1 pr 20	353CCP.2520	.144 x .328	34	-----	-----	679948
RG 59	1 pr 18	353CCP.2018	.199 x .465	54	H90224	252815	649948
RG 6	1 pr 18	353CCP.1818	.232 x .416	60	H90794	-----	639948
RG 6	1 pr 16	353CCP.1816	.221 x .430	64	H90061	-----	-----

Manufacturer's makes vary. Diameters (Col. B) and weights (Col. C) are conservative and for engineering purposes only. Consult manufacturer's data for individual make & model information.

Column A: Stranding (Number of Strands)
Column B: Approximate Cable Diameter in inches. (Cable sized to fit the connectors selected.)
Column C: Approximate - nominal cable weight in pounds per 1000ft

354SDR: 300V Specification 354: Serial-Digital-Interface Video Coax Cable SDI Coax
Solid BC / Foam PE Dielectric / 75Ω / Dual-Foil + 95% TC Braid Shield / PVC Jkt For Riser Use.

Conductor: Solid Bare Copper center conductor.

Insulation: Foam Polyethylene per UL 13 or UL 444 and as required for 75 Ohm Impedance, and so as to fit standard connectors.

Shielding: Dual Shield: Dual foil Alum.-Poly-Alum Tape bonded to the insulation + 95% TC Braid. .

Jacket: Black PVC Jacket. (Other colors as may be specified on drawings.)

Diameter: Per Table 354SDR - Column B.

Flame test: Passes UL 1666 flame test for cables installed in risers / CSA FT4.

Voltage: 150V or 300V rated. Cables listed as CL2R (only) are rated 150V. Others are rated 300V. Voltage rating is not printed on the jacket per UL 13 and UL 444

Application: Digital video including serial digital interface; HD video or precision analog video.

Installation: Suitable as open wiring in risers as Type CMR. Also suitable for use in horizontal runs since rated higher than those requirements.

Temp.: 0C to 60C; minimum range, continuous operating ambient temperature. .

QA: UL Listed as cable type CMR. Cable rating(s) shall be printed on the jacket.
UL file number will be printed on the jacket alongside AWG and cable type information.
UL label shall be displayed on every cable package near manufacturer's label.

Packaging: 1000ft or 500ft: spools, reels or boxes.

Appr. Manf: Manufacturer make and model shall be as selected from table 354SDR.

Availability: Cable shall be available from local distribution known to project management.

Table 354SDR: HD-SDI Coax; Sol BC / Foam PE / 75Ω / F+95% TC Braid / PVC Jkt. Riser Use

Const	AWG	SDN	B	C	TAPPAN	General Cable	Belden: WEST PENN	BELDEN
Mini RG 59	25	354SDR.25	.146	11	-----	-----	HD825	1281R
Mini RG 59	23	354SDR.23	.159	20	C90032	VDM230	-----	1855A
RG 59	20	354SDR.20	.242	37	C30114	395025	819	1505A
RG 6	18	354SDR.18	.275	48	C40193	395011	6350	1694A
RG 11*	14	354SDR.14	.405	92	C60013	395029	821	7731A

Manufacturer's makes vary. Diameters (Col. B) and weights (Col. C) are conservative and for engineering purposes only. Consult manufacturer's data for individual make & model information.

Column B: Approximate Cable Diameter in inches. (Cable sized to fit the connectors selected.)
Column C: Approximate - nominal cable weight in pounds per 1000ft

* Manufacturers listing as cable CL2R (only) – the voltage rating is 150V

354SDP: 300V Specification 354: Serial-Digital-Interface Video Coax Cable **SDI Coax**
Sol BC / Foam FEP Dielectric / 75Ω / Dual-Foil + 95% TC Braid Shield / Jacket **For Plenum Use.**

Conductor: Solid Bare Copper center conductor.

Insulation: Foam FEP per UL 13 or UL 444 and as required for 75 Ohm Impedance, and so as to fit standard connectors.

Shielding: Dual Shield: Dual foil Alum.-Poly-Alum Tape bonded to the insulation + 95% TC Braid; bond between tape and insulation as best as foam FEP is able to hold such a bond.

Jacket: White or natural LS-PVC Jacket. Or fluoropolymer jacket when necessary for the cable to pass the NFPA 262 flame test for use in plenum spaces. (Other colors as may be specified on drawings.)

Diameter: Per Table 354SDP - Column B.

Flame test: Passes NFPA 262 flame test for cables installed in plenum spaces / CSA FT6.

Voltage: 150V or 300V rated. Cables listed as CL2P (only) are rated 150V. Others are rated 300V. Voltage rating is not printed on the jacket per UL 13 and UL 444

Application: Digital video including serial digital interface; HD video or precision analog video.

Installation: Suitable as open wiring in plenum spaces as Type CMP. Also suitable for use in horizontal runs and in risers since rated higher than those requirements.

Temp.: 0C to 60C; minimum range, continuous operating ambient temperature. .

QA: UL Listed as cable type CMP or CL3P or CL2P. Cable rating(s) shall be printed on the jacket. UL file number will be printed on the jacket alongside AWG and cable type information. UL label shall be displayed on every cable package near manufacturer's label.

Packaging: 1000ft or 500ft: spools or reels.

Appr. Manf: Manufacturer make and model shall be as selected from table 354SDP.

Availability: Cable shall be available from local distribution known to project management.

Table 354SDP: HD-SDI Coax; Sol BC / Foam FEP / 75Ω / F+95% TC Braid / Appr. Jkt. Plenum Use

Const	AWG	SDN	B	C	TAPPAN	General Cable	Belden: WEST PENN	BELDEN
Mini RG 59	25	354SDP.25	.146	11	-----	-----	HD25825	1282P
Mini RG 59	23	354SDP.23	.159	21	D90033	VDM230TS	-----	1855P
RG 59	20	354SDP.20	.207	29	D90032	495023	25819	1506A
RG 6	18	354SDP.18	.236	39	D40061	495025	256350	1695A
RG 11*	14	354SDP.14	.351	83	D60007	495027	-----	7732A

Manufacturer's makes vary. Diameters (Col. B) and weights (Col. C) are conservative and for engineering purposes only. Consult manufacturer's data for individual make & model information.

Column B: Approximate Cable Diameter in inches. (Cable sized to fit the connectors selected.)
Column C: Approximate - nominal cable weight in pounds per 1000ft

* Manufacturers listing as cable CL2P (only) – the voltage rating is 150V

355RGBH: 300V Specification 355: Miniature High Resolution Coax Multiples **Mini RGBHV**
Miniature High Resolution Coax Multiples **For Horizontal Use.**
(many models for Riser Use.)

Conductor: Solid Copper (BC) center conductor, or stranded tinned copper center conductor.

Insulation: Foam Polyethylene per UL 13 or UL 444 and as required for 75 Ohm Impedance; and so that standard connectors fit.

Shielding: Dual shield; aluminum-polyester tape bonded to the dielectric, and 95% coverage of tinned copper braid, interlocked serve, or simple serve.

Coax Jkt: Individual coax PVC jacket. Thickness according to manufacturer's listing and such that standard connectors fit.

Color Code: Red, Green, Blue, White, Yellow, Black

Jacket: Black PVC Jacket.

Diameter: Per Table 355RGBH - Column B.

Flame test: Models rated CM, CL2, or CL3 passes UL 1685 Vertical Tray / CSA FT1.
Models rated CMR, CL2R, or CL3R pass UL 1666 / FT4.

Voltage: 150V or 300V rated. Cables listed as CL2 or CL2R (only) are rated 150V. Others are rated 300V. Voltage rating is not printed on the jacket per UL 13 and UL 444.

Application: Component RGB video, either analog or digital, component HDTV, computer gaming, computer animation production, long run HDMI between active conversion devices; classroom, meeting room, and auditorium video projectors; where BNC, RCA, for VGA connector(s) are provided..

Installation: Suitable open wiring / open cabling use in concealed horizontal runs as Type CM.
Models with a riser rating CL2R, CL3R, or CMR may also be used in risers.

Temp.: 0C to 60C; minimum range, continuous operating ambient temperature.

QA: UL Listed as cable type CM or CMR. Cable rating(s) shall be printed on the jacket.
UL file number will be printed on the jacket alongside AWG and cable type information.
UL label shall be displayed on every cable package near manufacturer's label.

Packaging: 1000ft or 500ft: spools or reels..
Appr. Manf: Manufacturer make and model shall be as selected from table 355RGBH.
Availability: Cable shall be available from local distribution known to project management.

Table 355RGBH: Miniature High Resolution Coax Multiples **For Horizontal Use.**

Const	SDN	B	C	TAPPAN	General Cable	Belden: WEST PENN	BELDEN
3 x 25 sol.	355RGBH.253	.325	44	H91925*	RGB250S*	-----	1277R*
4 x 25 sol.	355RGBH.254	.365	55	-----	RGBS250S*	-----	1278R*
5 x 25 sol.	355RGBH.255	.405	68	H91924*	RGBSC250S*	5CRGB	1279R*
6 x 25 sol.	355RGBH.256	.440	80	H92062*	RGBHVC250S*	-----	1280R*

Manufacturer's makes vary. Diameters (Col. B) and weights (Col. C) are conservative and for engineering purposes only. Consult manufacturer's data for individual make & model information.

Column B: Approximate Cable Diameter in inches. (Cable sized to fit the connectors selected.)
Column C: Approximate - nominal cable weight in pounds per 1000ft

* Model with an asterisk are rated for use in risers.

355RGBP: 300V Specification 355: Miniature High Resolution Coax Multiples Mini RGBHV
Miniature High Resolution Coax Multiples For Plenum Use.

Conductor: Solid Copper (BC) center conductor, or stranded tinned copper center conductor.

Insulation: Foam fluoropolymer per UL 13 or UL 444 and as required for 75 Ohm Impedance; and so that standard connectors fit.

Shielding: Dual shield; aluminum-polyester tape bonded to the dielectric, and 95% coverage of tinned copper braid, interlocked serve, or simple serve. Tape bond limited to the ability of fluoropolymer to hold such a bond.

Coax Jkt: Individual coax jacket. Thickness according to manufacturer's listing and such that standard connectors fit. Material such that the overall cable passes NFPA 262 flame test.

Color Code: Red, Green, Blue, White, Yellow, Black

Jacket: Black LS-PVC Jacket, or material as necessary for the overall cable to pass NFPA 262.

Diameter: Per Table 355RGBP - Column B.

Flame test: Passes NFPA 262 flame test for cables installed in plenum spaces / CSA FT6.
Voltage: 150V or 300V rated. Cables listed as CL2P (only) are rated 150V. Others are rated 300V. Voltage rating is not printed on the jacket per UL 13 and UL 444.

Application: Component RGB video, either analog or digital, component HDTV, computer gaming, computer animation production, long run HDMI between active conversion devices; classroom, meeting room, and auditorium video projectors; where BNC, RCA, for VGA connector(s) are provided..

Installation: Suitable open wiring / open cabling use in plenum spaces as Type CMP.
Also suitable for use as horizontal cable and in risers since rated higher.

Temp.: 0C to 60C; minimum range, continuous operating ambient temperature.

QA: UL Listed as cable type CMP. Cable rating(s) shall be printed on the jacket.
UL file number will be printed on the jacket alongside AWG and cable type information.
UL label shall be displayed on every cable package near manufacturer's label.

Packaging: 1000ft or 500ft: spools or reels..
Appr. Manf: Manufacturer make and model shall be as selected from table 355RGBP.
Availability: Cable shall be available from local distribution known to project management

Table 355RGBP: Miniature High Resolution Coax Multiples For Plenum Use.

Const	SDN	B	C	TAPPAN	General Cable	Belden: WEST PENN	BELDEN
3 x 26 str.	355RGBP.263	.260		-----	--------	WP258263	-----
5 x 26 str.	355RGBP.265	.364	74	H92042	RGBSC260TS	WP258265	-----
3 x 25 sol.	355RGBP.253	.313	49	H91946	--------	253CRGB	1277P
5 x 25 sol.	355RGBP.255	.330	72	H91930	--------	255CRGB	1279P
6 x 25 sol.	355RGBP.256	.369	80	-----	--------	-----	1280P

Manufacturer's makes vary. Diameters (Col. B) and weights (Col. C) are conservative and for engineering purposes only. Consult manufacturer's data for individual make & model information.

Column B: Approximate Cable Diameter in inches. (Cable sized to fit the connectors selected.)
Column C: Approximate - nominal cable weight in pounds per 1000ft

* Manufacturers listing as cable CL2P (only) – the voltage rating is 150V

356RFX: 300V Specification 356: Solid BC 50Ω RF Coax Cable
Solid Bare Copper / PE Dielectric / 50Ω / 95% Bare Copper Braid / PVC Jacket

50Ω RF Trans
For Limited Use.
(with models for horizontal use.)

Conductor: Solid Bare Copper. 356RFH.201B is solid tinned copper.

Insulation: Solid Polyethylene per UL 444 and as required for 50 Ohm Impedance and for standard Connectors to fit. Except 356RFH.201B is foamed polyethylene meeting requirements.

Shielding: 95% braid of Bare Copper or Tinned Copper.

Jacket: Black PVC Jacket. (Other colors as may be specified on drawings.)

Diameter: Per Table 356RFX - Column B.

Flame test: Passes UL 1685 Vertical Tray / CSA FT1 (applies to 356RFH.201.A and B)
Passes VW-1 flame test (applies to 356RFX.201C)

Voltage: 300V rated. Voltage rating is not printed on the jacket per UL 444.

Application: 50 Ohm RF transmission, wireless routers, access points, and microphones; citizen band, police, fire, and ambulance 2-way radio; land mobile radio, ship-to-shore marine and airport radio; in-building wireless and television broadcast transmission where TNC or N connector is provided.

Installation: 356RFH.201A and B may be used as open cabling in concealed horizontal runs as Type CM. 356RFX.20C for residential use or in conduit runs in commercial installation and 10 feet may be exposed.

Temp.: 0C to 60C; minimum range, continuous operating ambient temperature.

QA: UL Listed. Cable rating(s) shall be printed on the jacket.
UL file number will be printed on the jacket alongside AWG and cable type information.
UL label shall be displayed on every cable package near manufacturer's label.

Packaging: 1000ft or 500ft: spools, reels or boxes.

Appr. Manf: Manufacturer make and model shall be as selected from table 356RFX.

Availability: Cable shall be available from local distribution known to project management.

Table 356RFX: 50Ω RF Trans. Coax; BC / Sol. PE or FPE / BC Braid / PVC Jacket.

Const	SDN	A	Cond. Matl.	B	C	TAPPAN	Belden: WEST PENN	BELDEN
RG 58	356RFH.201A	1	BC	.193	36	C30084	-----	-----
	356RFH.201B	1	TC	.194	18	-----	813	-----
	356RFX.201C	1	BC	.193	25	-----	-----	8240

Consult manufacturer's data for individual make & model information.

Column A: Stranding (Number of Strands)
Column B: Approximate Cable Diameter in inches. (Cable sized to fit the connectors selected.)
Column C: Nominal cable weight in pounds per 1000ft. (Approximate.)

PART 3- EXECUTION

3.1 RELATED SECTIONS:

01. Sections listed in 1.2
02. 16100 Wiring Methods
03. 16121 Cable Enclosures
04. 16130 Raceway and Boxes
05. 16131 Conduits
06. 16132 Indoor service poles
07. 16134 Under floor raceway assemblies.
08. 16141 Wall, floor, and table boxes.
09. 16142 Wiring Devices: Power, audio, video, telecommunications and data devices.
10. 16150 Wiring Connections
11. 16155 Equipment Wiring

3.2 INSTALLATION OF RACEWAYS, BOXES, ENCLOSURES, AND EQUIPMENT RACKS

01. Where cable installation in raceway is indicated on project drawings, or is required due to cable type being inadequate for an open wiring run given the routing, or where best practice requires raceway: coordinate with the electrical contractor on raceway size, type, route and installation. Inform the electrical contractor as appropriate and necessary to ensure proper raceways are installed for the electronic wire and cable for correct and efficient operation of low voltage systems.
02. Where conduit systems are part of a UL listed Fire Resistive System such as for a "Circuit Integrity In Conduit" ("CIC") Fire Resistive System required to survive attack by fire: the conduit installation and it's supports system shall conform to the particular UL Fire Resistive System (FHIT) corresponding to the cable manufacturer make and model selected from Table 345UCIC or 345SCIC.
03. All dimensions locations, and conditions shall be verified in the field. The Contractor shall notify the Architect of any discrepancies before proceeding with the work
04. Communications cable, low voltage signaling and control wiring (any electronic low voltage cable) when installed in conduit or other enclosure or raceway, shall be installed in separate system of conduit, enclosure, or other raceway than electric power wire and cable. Where raceway provides separate partitions designed to accommodate end-to-end separate pathways and these provide adequate separation so there is no malfunctioning their use is permitted when manufacturer's instructions are followed.
05. Fire alarm circuit wiring and control wiring associated with the fire alarm system devices shall be installed in a dedicated raceway system. This system shall not be used for any other wire or cable.
06. Installation of other low voltage systems shall be in separate dedicated pathways for each system where it is specified or where it is customary or where systems are bid and awarded separately.
07. Observe pull box sizing, spacing, and number of bends between pull boxes specified in ANSI/TIA/EIA 569-A.

08. Install long radius elbow fittings or sweeps everywhere space allows for their use. Minimum radius used shall be 6 inches, or 10x the cable diameter.

09. Raceway shall be completely and properly installed and inspected before cable is installed.

10. Provide fire stopping at all fire rated wall, floor, and partition penetrations with UL listed firestopping materials or systems as called for by NFPA 70/NEC.

11. Pull boxes and any unavoidable junctions boxes (splices are not allowed) shall be attached to structural members in a location that will be subsequently accessible. If location would not be accessible otherwise, an access door shall be provided.

12. Equipment enclosures shall be installed at the locations indicated and ventilated as required to provide conditions specified by the manufacturers of the electronic equipment therein.

13. All pull boxes and any junction boxes shall be labeled. Labels shall state the system(s) therein and cabling used for each accessible system whether connected or passing by. Labeling shall be indelible to contact with water and oil. In correctional facilities labeling that would be visible to inmates quarters shall be subject to approval by architect, engineer, or other owner's designated representative.

3.3 INSTALLATION OF LOW VOLTAGE ELECTRONIC CABLE

01. Cable to all systems' devices is included in the projects scope, including cabling which may not appear on drawings. All sufficient and necessary low voltage cabling for full and proper functioning of access control, security and safety systems, audio, video, building management, and communications systems specified and included in the project scope – including cabling runs that may not appear on project drawings -- shall be provided and installed at no additional cost above the base project bid.

02. Comply with paragraph 1.7 02 if cable has been stored in a cold location. Use of heat lamps or forced air heating applied directly onto cable is not permitted as a means to condition it for installation.

03. Cabling shall meet the requirements of NFPA 70 Articles 725, 760, and 800 as applicable.

04. Comply with ANSI/EIA/TIA 569-A Commercial Building Standard for Telecommunications Pathways and Spaces – including all addendums – as applicable.

05. Comply with BICSI TDDM; <u>Telecommunications Distribution Methods Manual</u>, as applicable.

06. Comply with NECA 1 as applicable.

07. Comply with NFPA 72 for Fire Alarm circuit cabling for Fire Alarm cabling installation.

08. Fire Alarm system cabling shall have jacket color(s) and/or striping different than other systems and unique within the building. Striped jackets will have factory original extruded color stripe. Painted-on striping is not permitted.

09. Do not install cable with damaged jacket.

10. Comply with best practices prescribed in training manuals published by InfoComm, NSCA, NICET or BISCI as applicable to installer's corresponding certification.

11. Always observe maximum pulling tension and minimum bending radius recommendations and conventions according to cable type.

12. Low voltage cable for remote control signaling, audio, video, control, and communications shall be installed in separate pathways than cable and wire for electrical power. Where raceways are required they shall be completely and properly installed and inspected according to section 3.2 before cable is installed therein.

13. Splicing is not permitted unless approved. Coaxial cable may be spliced with appropriate fittings. Fire Alarm and Access Control circuits may be spliced with approval. Circuit Integrity Fire Alarm circuits for systems required to survive attack by fire shall not be spliced within the fire zones where their survivability is required for the system to function as designed.

14. Take customary precautions and provide customary protections as needed to avoid electrical interference with low voltage circuits.

15. Provide spacing as indicated below from lighting fixtures and other electrical devices.
    A. Separation between Low Voltage cables in grounded metallic raceways and unshielded power lines or electrical equipment:
        a. 2.5 inches from electrical equipment or lines rated less than 2kVA
        b. 6.0 inches from electrical equipment or lines rated 2kVA – 5kVA
        c. 12 inches from electrical equipment or lines rated higher than 5kVA
    B. Separation between grounded metallic raceways for Low Voltage cables and other grounded metallic raceways for power lines:
        a. 3.0 inches from conduit for lines rated 2kVA – 5kVA
        b. 6.0 inches from conduit for lines rated higher than 5kVA
    C. Maintain 48 inches between Low Voltage cables and motors, transformers, and electrical switchgear.
    D. Maintain 5 inches between Low Voltage cables and fluorescent fixtures.

16. Where low voltage cable must cross over or under electrical lines maintain 90 degree perpendicular crossing, as well as the customary separation.

3.3.1 INSTALLATION OF OPEN-WIRING

01. Cabling shall meet the requirements of NFPA 70 Articles 725, 760, and 800 as applicable.

02. Cabling for Fire Alarm systems shall meet the requirements of NFPA 72.

03. Comply with ANSI/EIA/TIA 569-A Commercial Building Standard for Telecommunications Pathways and Spaces – including all addendums – as applicable.

04. Do not install cable with damaged jacket.

05. Comply with best practices prescribed in training manuals published by InfoComm, NSCA, NICET or BISCI as applicable to installer's corresponding certification.

06. Splicing is not permitted unless approved. Coaxial cable may be spliced with appropriate fittings. Fire Alarm and Access Control circuits may be spliced with approval. Circuit Integrity Fire Alarm circuits for systems required to survive attack by fire shall not be spliced within the fire zones where their survivability is required for the system to function as designed.

07. Take customary precautions and provide customary protections as needed to avoid electrical interference with low voltage circuits.

08. Provide spacing as indicated below from lighting fixtures and other electrical devices.
    A. Separation between Low Voltage cables and power lines or electrical equipment:
        a. 6 inches from electrical equipment or lines rated less than 2kVA
        b. 12 inches from electrical equipment or lines rated 2kVA – 5kVA
        c. 18 inches from electrical equipment or lines rated higher than 5kVA
    B. Maintain 48 inches between Low Voltage cables and motors, transformers, and electrical switchgear.
    C. Maintain 5 inches between Low Voltage cables and fluorescent fixtures.

09. Where low voltage cable must cross over or under electrical lines maintain 90 degree perpendicular crossing, as well as the customary separation.

10. Install cabling with horizontal cable guides and vertical cable guides in respective routings in telecommunications spaces.

11. Cable shall not be in contact with pipes, plumbing, ductwork or other potentially damaging items.

12. Cable shall not be run through structural members except where members have designed pass-throughs and where such routing is pre-approved and where appropriate protection from any sharp edge is provided.

13. Suspend cables run as open wiring a minimum of 8 inches above ceilings using cable supports spaced 30 inches apart maximum for Part 2 Specifications 33, 35, and 36 (UTP; EIA/TIA 568B Category 3, Category 5e, and Category 6 respectively) and 60 inches apart maximum for others.

14. Support Part 2 Specifications 33, 35, and 36 (UTP; EIA/TIA 568B Category 3, Category 5e, and Category 6 respectively) 6 inches maximum from boxes, fittings, outlets, cabinets, racks, frames and terminus.

15. Where cable is run as open wiring above a ceiling: cable shall be plenum rated.

16. Where cable is routed exposed under a raised floor; cable shall be plenum rated.

17. Plenum rated cable may be installed in all indoor low voltage cable routings. In building spaces designated as plenum space or air-handling space, plenum rated cable is required.

18. Raised flooring system shall be fully installed and inspected prior to installation of cable routed below the raised floor.

19. Provide 144 inches (12 feet) of cable neatly coiled into a 12 inch diameter coil at each feed point where cable is routed.

3.4 CONNECTIONS

01. Comply with the requirements of Section 13700 Security Access and Surveillance for connecting, terminating, and identifying cabling for systems covered therein.

02. Comply with the requirements of Section 13710 Intrusion Detection for connecting, terminating, and identifying cabling for systems covered therein.

03. Comply with the requirements of Section 13800 Building Automation and Control for connecting, terminating, and identifying cabling for systems covered therein.

04. Comply with the requirements of Section 13851 Fire Detection and Alarm for connecting, terminating, and identifying cabling for systems covered therein including Digital, Addressable Fire Alarm System.

05. Comply with the requirements of Section 16100 Basic Electrical Materials and Methods as applicable.

06. Comply with the requirements of Section 16700 Communications, 16710 Communications Circuits, and 16770 Communications Transmission and Reception Equipment for connecting, terminating, and identifying cabling for systems covered therein.

07. Where Category 3, Category 5e or Category 6 cable is indicated and RJ-45 jacks are indicated; comply with TIA/EIA 568-B.1

08. Install 110 style IDC termination hardware where Category 3, Category 5e, or Category 6 cable is otherwise terminated.

09. Observe best practices detailed in BICSI training manuals Cable Termination Practices.
10. Comply with the requirements of Section 16810 Sound and Video Circuits for connecting, terminating, and identifying cabling for systems covered therein.
11. Install 75 Ohm coaxial connectors and other fittings; maintaining 75 Ohm system impedance wherever 75 Ohm coaxial cable is specified and installed.
12. Install 50 Ohm coaxial connectors and other fittings; maintaining 50 Ohm system impedance wherever 50 Ohm coaxial cable is specified and installed
13. Use appropriate tools to ensure proper connections.
14. Bundle, lace, and train cabling to terminus points with no excess.
15. Cabling terminated in any enclosure associated with a fire alarm system shall be connected to terminal blocks. Connect using approved plug connectors, pressure-type terminal blocks, or crimp-on spade plugs.
16. Mark terminations according to the system's wiring diagrams.
17. Where terminations are in high density configurations - use numbered terminal strips.

3.5 FIRE STOPPING

01. Provide fire stopping at all fire rated wall, floor, and partition penetrations with UL listed fire stopping materials or systems as called for by NFPA 70/NEC.
02. Comply with requirements of Division 07840 "Fire Stopping."
03. Comply with TIA/EIA 569-A Annex A Fire Stopping.

3.6 GROUNDING AND SURGE PROTECTION

01. Comply with the requirements of Section 16060 Grounding and Bonding.
02. Comply with ANSI/TIA/EIA- 607 Grounding and Bonding Requirements for Telecommunications in Commercial Buildings.
03. Comply with the Grounding and Bonding requirements in BICSI TDDM; Telecommunications Distribution Methods Manual.
04. Provide singlepoint grounding for individual systems as recommended by system device manufacturers and IEEE.
05. Provide surge protection and clamping for all circuits.
06. Coordinate all grounding, bonding, surge protection, and clamping of systems' circuits with respective system device manufacturers' manuals and personnel.
07. Coordinate grounding and surge protection requirements with other trades and system integrators so no ground loops will be closed. Provide isolation transformers and other equipment as required.
08. Provide and install all cabling and insulation materials for proper grounding and bonding of low voltage systems, and supports for same, as required.

3.7 IDENTIFICATION

01. Comply with the requirements of Section 16075 “Electrical Identification” as applicable.

02. Comply with ANSI/TIA/EIA 606-A Administration Standard for the Telecommunications Infrastructure of Commercial Buildings

3.8 FIELD QUALITY CONTROL

01. Inspect Electronics Wire and Cable upon arrival at the project in compliance with paragraph 1.7 01. of this section.
02. Test and Inspect all systems to ensure their performance as warranted.
03. Test for opens and shorts and correct any found.
04. Visually inspect all system cabling for deformities, evidence of strain, or bends or lower radius than the minimum recommended.
05. End-to-end verification testing is required for LAN systems where Category 5e and Category 6 cable types are specified and installed.

END SECTION 13410

PUBLIC CABLE™ 1.0

TABLE OF CONTENTS

MASTER TABLE OF PUBLIC CABLE™ 1.0 PRODUCTS

DETAIL TABLE OF PRODUCTS – Stranded Bare Copper, UL minimum Insulation

Solid Bare Copper, UL minimum Insulation

Stranded Tinned Copper, Upgrade Insulation

INDEX

2

9

A

B

C

D

Q

R

CPSIA information can be obtained at www.ICGtesting.com
Printed in the USA
LVOW02s0354160913
352576LV00002B/13/P

9 780988 236400